Recent Research in Greenhouse Gases

Editor: Joshua Garrett

R CALLISTO
REFERENCE
www.callistoreference.com

Callisto Reference,
118-35 Queens Blvd., Suite 400,
Forest Hills, NY 11375, USA

Visit us on the World Wide Web at:
www.callistoreference.com

ISBN: 978-1-64116-188-6 (Hardback)

Cataloging-in-Publication Data

Recent research in greenhouse gases / edited by Joshua Garrett.
    p. cm.
Includes bibliographical references and index.
ISBN 978-1-64116-188-6
1. Greenhouse gases. 2. Greenhouse gas mitigation. I. Garrett, Joshua.
TD885.5.G73 R43 2019
363.738 74--dc23

Table of Contents

Preface

This book was inspired by the evolution of our times; to answer the curiosity of inquisitive minds. Many developments have occurred across the globe in the recent past which has transformed the progress in the field.

Greenhouse gases emit and absorb radiant energy, in the thermal infrared zone. Gases like carbon dioxide, nitrous oxide, chlorofluorocarbons, methane, hydrofluorocarbons, ozone and water vapor are the prominent types of greenhouse gases which result in an increase in temperature of the Earth's surface. This effect is called the greenhouse effect. Some gases have an indirect radiative effect. They may break down in the atmosphere producing a greenhouse gas, or undergo chemical reactions in the atmosphere thereby changing the concentrations of greenhouse gases. Such indirect effects along with the characteristics of each atmospheric gas species and their abundance have a significant impact on the greenhouse effect. Some of these characteristics are the atmospheric lifetime, global warming potential and radiative forcing. Elevated levels of greenhouse gases can potentially lead to rising sea levels, extreme weather phenomena, negative effects on agricultural productivity and loss of biodiversity. This book aims to shed light on some of the unexplored aspects of greenhouse gases and the recent researches in this field. It includes contributions of experts and scientists, which will provide innovative insights into this subject. It will serve as a valuable source of reference for graduate and postgraduate students as well as for experts.

This book was developed from a mere concept to drafts to chapters and finally compiled together as a complete text to benefit the readers across all nations. To ensure the quality of the content we instilled two significant steps in our procedure. The first was to appoint an editorial team that would verify the data and statistics provided in the book and also select the most appropriate and valuable contributions from the plentiful contributions we received from authors worldwide. The next step was to appoint an expert of the topic as the Editor-in-Chief, who would head the project and finally make the necessary amendments and modifications to make the text reader-friendly. I was then commissioned to examine all the material to present the topics in the most comprehensible and productive format.

I would like to take this opportunity to thank all the contributing authors who were supportive enough to contribute their time and knowledge to this project. I also wish to convey my regards to my family who have been extremely supportive during the entire project.

Editor

Greenhouse Gases Production from Some Crops Growing Under Greenhouse Conditions

Fernando López-Valdez, Fabián Fernández-Luqueño,
Carolina Pérez-Morales and Mariana Miranda-Arámbula

Additional information is available at the end of the chapter

Abstract

Greenhouse gases, such as carbon dioxide (CO_2), nitrous oxide (N_2O) and methane (CH_4), play an important role in global climate change. For example, CO_2 production occurs as a result of the seasonal cycles of the biotic processes of photosynthesis and respiration, as well as through anthropogenic activities and abiotic processes such as the burning of fossil fuels. Many activities, such as Agribusiness (the production of crops and animals for food) create greenhouse gases. Our research group has studied several soil treatments such as wastewater, wastewater sludge, vermicomposting, and urea among others, in order to study the effects of soil treatments on the production of greenhouse gases (CO_2, N_2O, and CH_4) in several cultivars, but mainly in maize, sunflower and the common bean. The principal aim of this chapter is to show how these greenhouse gases are affected by the type of treatment, the properties of the soil, and the cultivar in question. We also look at which processes are involved in the production of CO_2, N_2O, and CH_4 from cultivated soil. We present a review of several experiments carried out under *in vitro* or greenhouse conditions.

Keywords: Greenhouse gases production, wastewater sludge, fertilizers, treatments

1. Introduction

Global food demand is increasing rapidly, while the associated potential negative environmental impacts are also growing. Land clearance, the intensive use of existing croplands, inadequate agricultural management systems, and soil pollution could all contribute to an increase in the production of greenhouse gases (GHG). Understanding the future environmental impacts of global crop production, while at the same time achieving greater yields with lower impacts, requires quantitative assessments of future crop demand and an understanding of how different production practices affect yields and environmental variables [1].

It is well known that crop management systems, the quality of the soil and the weathering conditions are just some of the factors used in order to assess production of GHG. Therefore, an understanding of the future environmental impacts of crop production is essential in order to achieve greater crop yields without decreasing the quality of the environment and social welfare. Additionally, Tilman [2] reported that the recent intensification of agriculture, coupled with the prospect of even further intensification in the future, will have major detrimental impacts on the world's ecosystems.

Agriculture is rightly recognized as a source of GHG production, with concomitant opportunities for its mitigation. In fact, agricultural soils can constitute either a net source or sink of the three principal GHG [3]. Soil management practices can influence GHG flux by changing at least one of the following soil properties and its associated management: 1) The soil climate (temperature and water content); 2) The physicochemical environment of the soil; 3) The soil's microorganisms (diversity and abundance); 4) The amount and chemical composition of organic or mineral fertilizers applied to the soil; and 5) Pesticides might have a strong effect on the soil microbiota (type and amount). Even a minimal change in one or more of the properties described above could control the rate and extent of GHG production and also affect the aeration and diffusion of these gases.

The objective of this chapter is to discuss how the soil production rate of GHG is affected by treatment type, soil properties, and cultivar. This review will also discuss which processes are involved in the production of CO_2, N_2O, and CH_4 when crops are grown under *in vitro* or greenhouse conditions, and will discuss how these processes work.

2. The Atmosphere, Global Climate Change and Greenhouse Gases

The atmosphere of the Earth has evolved and changed over time and had reached a point of equilibrium. However, anthropogenic activities arising from the Industrial Revolution and subsequent development have changed Earth's atmospheric conditions. Since the industrial era began, a new phenomenon has been observed — that of global climate change (GCC). Many different sources are involved in the production of gases, with concern centering on the production of GHG in particular, as these gases are the ones implicated in the increasing rate of global warming on Earth. The main gases involved in this warming are methane (CH_4), nitrous oxide (N_2O), and carbon dioxide (CO_2). Additional and important GHGs include water vapour, which has an effect on global climate change that can be used as a baseline with which to compare the remaining GHG. The production of these gases arises as a result of anthropogenic activities, mainly the combustion of fossil fuels (CO_2), extensive livestock and cattle farming (CH_4), and agriculture (N_2O) through denitrification or nitrification processes, and occasionally CO_2 depending on the type of fertilization employed.

As we can see, global climate change is a phenomenon caused by GHG that are emitted into the atmosphere. However, the main problem is not the emission of these gases, because these gases have actually been present in the Earth's atmosphere for thousands of years and they are the products of natural processes such as volcanic eruptions, plant and animal respiration and the microbial decomposition of organic matter. The contribution of human activity has resulted in the production of large amounts of these gases and their increased concentration in the atmosphere results in global warming. The most obvious effects of global warming are the continuous increase of global temperature and the changes in atmospheric conditions. All the elements of the environment are interrelated, and as a consequence, changes in one of them lead to changes in others. Sometimes these changes are small and imperceptible, while others can be very obvious. The rate of these changes is very important because if they are too rapid, then the ability of organisms to adapt to the new conditions might not be sufficient to ensure their survival as the natural process of adaptation takes thousands and thousands of years as a part of the evolutionary processes of life on Earth. The effects of these phenomena are the extinction of species and other serious negative effects on both the agriculture and fishing industries that are important economic activities the world over.

It is important to mention that the likely impacts of global warming could be different in different types of ecosystems because of the difference in climatic conditions in those ecosystems, but the effects on the abundance and distribution of biodiversity will be constant. All of these facts suggest that the natural conditions of the planet are being seriously affected by global climate change, global warming and GHG production, so we have a serious and worldwide environmental problem to address. However, there are many strategies, such as the use of alternative sources of energy, which could be implemented around the world to mitigate the damage being caused to the planet and to promote environmental awareness with favorable results in the future.

3. Experiments under *in vitro* conditions

We are interested in understanding which soil processes are involved in GHG production, and how they work, in several treatments or fertilizers. The production of CO_2, N_2O, and CH_4 when crops were grown under *in vitro* or greenhouse conditions was studied. These experiments were carried out using different types of soil (nitrogen depleted and/or alkaline-saline) and several crops were studied.

One of the experiments was conducted in order to investigate the evolution of nitrogen and its loss as a part of the nitrogen cycle. Different fertilizers or treatments were tested. These were ammonium sulphate [$(NH_4)_2SO_4$, 200 mg NH_4^+ kg^{-1}], wastewater sludge (200 mg NH_4^+ kg^{-1}), sterile wastewater sludge (200 mg NH_4^+ kg^{-1}) and a control (distilled H_2O). All of the treatments were added with KNO_3 at 100 mg N kg^{-1}, in two different soils (one agricultural and N depleted soil, and the second a saline-alkaline and N depleted soil) at 40% of water holding capacity (WHC) under *in vitro* conditions for 56 days. The variables were CO_2, N_2O, NH_4^+-N, NO_2^--N, and NO_3^--N, and were measured and assessed [4]. The soils used were specifics, i.e. the first soil was an agricultural soil which had mainly been cultivated with maize (> 25 years), low fertilization (< 50 kg N ha^{-1}) and was both C and N depleted (6.5 g organic C kg^{-1}, 0.2 g total N

Kjeldahl kg^{-1}, pH 7.8, electrolytic conductivity (EC) 1.0 dS m^{-1}, and the textural soil classification was loamy sand) from Otumba, in the State of Mexico (Mexico) (19° 42′ N, 98° 49′ W). The second soil was classified as an uncultivated soil (some grasses and small trees could be found) as a result of its former lake bed origin. It was found to be N depleted and alkaline-saline, pH 10.3, EC 12.4 dS m^{-1}, 49 g organic C kg^{-1}, and 0.6 total N Kjeldahl kg^{-1}, from Texcoco, State of Mexico (Mexico) (19° 30′ N, 98° 53′ W). The results showed that production of CO_2 from the Otumba soil was not affected by the addition of NH_4^+ or NO_3^-, i.e. both fertilizers produced a similar amount of CO_2, approximately 350 mg CO_2-C kg^{-1} dry soil. The sterilized sewage sludge increased the production of CO_2, > 1,000 mg CO_2-C kg^{-1} dry soil, i.e. over twice the production compared with that of the controls (soils treated with ammonium or nitrate). When wastewater sludge was added, the CO_2 production was ~ 3,100 mg CO_2-C kg^{-1} dry soil, a value twice that of sterilized sludge and eight times that of the controls. In the Texcoco soil, a similar contour was found in the CO_2-C dynamics. The control treatments present > 350 mg CO_2-C kg^{-1} dry soil. The soil treated with sterilized sludge showed a concentration of CO_2 that was > 1,100 mg CO_2-C kg^{-1} dry soil (2.1 times that of the controls) while the treatment with sewage sludge was ~ 2,100 mg CO_2-C kg^{-1} dry soil (over five times that of the controls). Different soils showed similar contours in C dynamics, and these results revealed that CO_2 production was not particularly different and that the processes involved were rather similar in the different soils investigated. Wastewater sludge is characterized by large amounts of organic matter (organic C) and it's suppose got a large amounts of organic matter easily decomposable. So, the sludge is easily and rapidly mineralized in both soils [4]. When sewage sludge was sterilized, the microbiota were destroyed and this might have affected the properties of the organic matter in the sludge [4]. It could be that the organic matter was more readily accessible for the soil microorganisms, so that the production of CO_2 should be higher in soils with sterilized sludge, but the results revealed that this is not necessarily true. The results also showed that the soil microorganisms and the sludge microorganisms could be exerting a synergistic action on the degradation of organic matter because the degradative action of the microorganisms of the soil or the sludge alone cannot improve the degradation of organic matter in the treated soil or in the sludge [4].

The ammonium dynamics showed that the initial concentrations of N were reduced after the first 3 days, and after that, a release of the mineral occurred from day 3 up to day 14. Later still, the concentration of ammonium decreased by up to < 14 mg N kg^{-1} dry soil for all the treatments in both the Otumba and Texcoco soils, and the ammonium concentration decreased by up to < 2 mg N kg^{-1} dry soil for all treatments, except for the soil treated with sterilized sludge, < 31 mg N kg^{-1} dry soil. The contour of the ammonium dynamics was similar in both the Otumba and Texcoco soils. Many abiotic and biotic processes might affect the concentration of NH_4^+ in soil, such as NH_4^+ fixation in the soil matrix, volatilisation of NH_3, and immobilization or oxidation of NH_4^+. Some soil processes were occurring at too low a level to be detectable, such as NH_4^+ fixation and the volatilisation of NH_3. The nitrate dynamics were similar in both soils. The concentration of NO_3^- was ~120 mg N kg^{-1} dry soil in the control treatment in both soils. The ammonium concentration was similar in both soils, > 200 mg N kg^{-1} dry soil, treatments with sludge reached > 255 mg N kg^{-1} dry soil and > 300 mg N kg^{-1} dry soil in the Texcoco and Otumba soils respectively, and soils treated with sterilized sludge increased the concentration to > 300 mg N kg^{-1} dry soil in the Texcoco soil, while in the Otumba soil it was > 325 mg N kg^{-1} dry soil. These results suggest that soil fertilized with wastewater sludge showed an increased NO_3^- concentration with a hypothetic mineralisation of ~ 60% at day 56 [4].

The production of N_2O was measured in both soils for seven days under C_2H_2 (10% v/v) conditions. The control treatment showed a N_2O production of < 0.02 mg N kg^{-1} with or without C_2H_2 in the Otumba soil. N_2O production increased when ammonium was added to the Otumba soil, 0.04 mg N kg^{-1}, but the addition of C_2H_2 reduced it to 0.01 mg N kg^{-1}. When the sterilized and non-sterilized sludge were added to the Otumba soil, N_2O production increased to 1 mg N kg^{-1} dry soil without C_2H_2 and 0.49 mg N kg^{-1} dry soil with C_2H_2. In the Texcoco soil, the control treatment was below 0.005 mg N kg^{-1} dry soil, but when C_2H_2 was added an increase to 0.09 mg N kg^{-1} dry soil was observed. Soil treated with ammonium increased the production of N_2O (0.04 mg N kg^{-1}), but under C_2H_2 conditions the concentration was low at 0.01 mg N kg^{-1} dry soil when compared with the control treatment. Soils treated with sterilized and non-sterilized sludge increased the production of N_2O (2.1 mg N kg^{-1} and 0.75 mg N kg^{-1}) compared with the control treatment, however, the addition of C_2H_2 increased the N_2O concentration of soil treated with sludge (2.1 mg N kg^{-1} dry soil) compared with soil treated with sterilized sludge (1.8 mg N kg^{-1}). It could be argued that the addition of easily decomposable organic matter into the soil will induce the denitrification process, where NO_3^- is reduced to N_2O and N_2 as final products. There are factors that could be important in controlling the production of N_2O in denitrification, such as oxygen, pH and the ratio of nitrate / available carbon [5]. There are additional parallel factors for NO that are less well understood [5]. It was evident that in a soil treatment of wastewater sludge or sterilized sludge, the N loss was increased. The microorganisms from the soil and the sludge were found to be working together in order to degrade the organic matter in the sludge. In addition, it could be suggested that more denitrifiers may be present in the Texcoco soil than in the Otumba soil, and a decrease of 50 mg NO_3^--N was counted at day 56, and the N_2O concentration was approximately double (plus 1 mg N_2O kg^{-1} dry soil) that of the Otumba soil on the day 7 [4]. When NO_3^- (an e$^-$ acceptor) is present in excess compared to organic C (an e$^-$ donor), the denitrifiers could be said to be "spendthrift" with respect to NO_3^- and in general produce N_2O as the major product. When the same NO_3^- is limited, the denitrifiers use it to its maximum potential as an e$^-$ acceptor and reduce it all to N_2 (dinitrogen) [5]. Schimel and Holland (2005) explain that while the major producer of NO is nitrification, N_2O can also be produced in large quantities by nitrification or denitrification due to the fact that it is less reactive and can outflow from soils — even wet soils.

In our study, the microorganisms in the soil and sludge acted together synergistically in the reduction process, i.e. N_2O to N_2, the denitrification process, and in the Texcoco soil under C_2H_2 conditions, but the main contribution was from the soil microorganisms under the same conditions. In particular, the production of N_2 was almost 50% of the total gas evolved. When sterilized sludge was applied to the Texcoco soil, minimal N_2 was produced (~0.2 mg N_2O-N kg^{-1}). Furthermore, untreated soil showed an increase in N_2O when C_2H_2 was added (~0.07 mg N_2O-N kg^{-1}) and the N_2 produced was approximately 75% of the total gas evolved, under the conditions established. In the Otumba soil, the N_2O was produced by nitrification process (at

40% WHC), and showed in the soil treated with ammonium, sludge and sterilized sludge at 0.04, 1.0, and 1.0 mg N_2O-N kg^{-1} respectively, when compared with untreated soil. N_2O production was low when C_2H_2 was added to the Otumba soil, and no N_2 was produced under these conditions. It can be established that when organic matter with a high N content is added to soil, it significantly increases N_2O production compared to untreated soil or soil fertilized with ammonium [$(NH_4)_2SO_4$] in both soils (an ordinary soil and an alkaline-saline soil). The major source of production of N_2O was found in all treated soils to be as a result of the nitrification process, and the production of N_2 was not recorded in the Otumba soil in this experiment. In the Texcoco soil, the major source of production of N_2O was as a result of the denitrification process by microorganisms in the soil, and the production of N_2 was approximately 50% of the total gas evolved (1.1 mg N_2O-N kg^{-1}).

4. Experiments under greenhouse conditions

4.1. First experiment under greenhouse conditions

Subsequent studies were established in order to better understand plant growth and the production of GHG (CO_2 and N_2O) when a regular *Bacillus subtilis* strain was inoculated on the surface of the sunflower (*Helianthus annuus* L.) cultivar seeds under greenhouse conditions. The *B. subtilis* strain was characterized as PGPR, i.e. showing antagonistic activity against *Fusarium oxysporum* and *Rhizoctonia solani* AG1, phosphate solubilizing activity, 1-aminocy-clopropane-1-carboxilate deaminase, and indole-3-acetic acid production. The strain was found as regular PGPR, for more details see [6].

The soil was collected from Alcholoya (Acatlán, in the State of Hidalgo, Mexico). This soil is an agricultural soil with a pH of 6.5, electrolytic conductivity (EC) 0.7 dS m^{-1}, 846 g kg^{-1}, organic C content was 11.1 g C kg^{-1} soil, and total N content 1.0 g N kg^{-1} soil. The soil was sampled from three different plots (400 m^2), ~ 800 kg was obtained and each plot was pooled separately and passed through a 5 mm sieve. Thirty-six sub-samples of 6.5 kg of soil from each plot were placed in cylindrical pots ($\varnothing$ = 16 cm, 50 cm) with 7 cm of gravel in the bottom. Four treatments were applied, with nine pots for each of the three soil sites sampled ($n = 27$). The first treatment was unfertilized and uncultivated soil (used as the CONTROL treatment), the second treatment was unfertilized soil cultivated with sunflowers (SUNFLOWER treatment), the third treatment was soil cultivated with sunflowers and fertilized with 0.5 g urea (75 kg N ha^{-1}, UREA treatment), and the fourth treatment was soil cultivated with sunflowers (seeds were dressed with *B. subtilis*) and fertilized with 0.5 g urea (BS treatment). All treatments were irrigated with tap water, with an additional input of 19 kg mineral-N ha^{-1} as NO_2^--N and NO_3^--N. In addition, twelve days after the emergence of the plantlets, they were fertilized with another 0.5 g of urea (the UREA and BS treatments), giving a total amount of 150 kg N ha^{-1}; and three weeks after sowing the plantlets were drenched with 4 mL of a bacterial suspension (at the same concentration as described above) adjacent to the plantlet roots at a depth of 3 cm. From the beginning of the experiments and approximately every two days for the following 30 days, the pots were closed air-tight and their atmosphere was analysed for CO_2 and N_2O at times of 0, 3, 15 and 30 mins. The experiment was repeated twice [6].

The daily CO_2 production rate for some treatments was large at the beginning of the experiments (data not shown). The daily CO_2 production rate showed a drop, remaining < 5 mg C kg^{-1} day^{-1} on day 2 and after that it remained at < 8 mg C kg^{-1} day^{-1} for all treatments up to end of the experiments. The mean CO_2 production rate was not significantly different between treatments. The daily N_2O production rate remained ≤ 0.75 µg N kg^{-1} day^{-1} for the SUNFLOWER and CONTROL treatments. Meanwhile, in the BS and UREA treatments, the daily N_2O production rate remained ≤ 2.1 µg N kg^{-1} day^{-1} with a maximum score of production in the first 14 days. The mean of the N_2O production rate of BS treatment was significantly high when compared with the SUNFLOWER and CONTROL treatments. Cultivating soil with sunflowers (SUNFLOWER treatment) did not affect the production of CO_2 compared with the CONTROL treatment (uncultivated soil). It is well known that cultivated soil frequently increases the production of CO_2, possibly due to the activities of the microorganisms degrading the easily decomposable organic matter such as the dying roots and root exudates in the rhizosphere, thereby increasing the production of CO_2 from the soil. Soils cultivated with sunflowers and fertilized with urea did not affect the production of CO_2. Applying urea to soil commonly has no effect on the production of CO_2 from soils. However, the production of CO_2 might be stimulated when urea is applied to N depleted soil as reported in Phillips and Podrebarac (2009), where the CO_2 production was tripled when 112 kg urea-N kg^{-1} was applied to an arable soil. Increases in CO_2 production with several doses of urea-N application indicate that agronomic-scale N inputs might stimulate microbial carbon cycling in arable soils [7]. The inoculation of *B. subtilis* on sunflower roots in soils fertilized with urea did not show any effect on the production of CO_2 compared with the UREA treatment. The application of UREA to the soil resulted in more than a doubling of the mean N_2O production rate when compared with the CONTROL treatment. The production of N_2O and NO in the soil is the result of factors such as ion concentrations and soil conditions under chemical disequilibrium, i.e. the oxidative process of NH_4^+ to NO_3^- under aerobic conditions, nitrification, and a reductive process of NO_3^- to N_2 under anaerobic conditions, and denitrification or the nitrifier denitrification [5, 8, 9]. According to the IPCC, N_2O is the main gas produced and released to the atmosphere by soil microorganisms [9]. So, when urea is applied into the soil, hydrolysis of the urea is immediately started, releasing NH_4^+ into the soil, which is rapidly transformed to NO_3^- with the simultaneous production of N_2O by the same process. A high concentration of NO_3^- in soil favours the production of N_2O due to the presence of anaerobic micro-sites in highly compacted soil.

A principal component analysis (PCA) was undertaken to investigate several plant biometric parameters and soil properties in the production of N_2O and CO_2. The analysis was carried out to include all variables. The PCA revealed that BS treatment has an effect on the shoots of plants, i.e. the shoot length, dry weight of shoot, and the fresh weight of shoot. UREA treatment has an effect on roots, i.e. the dry weight of roots, length of roots, and fresh weight of roots, and a minimal effect on seed weight. Otherwise, the soil properties PCA showed that both the BS and UREA treatments had an effect on NO_3^- and EC at depths of 0-15 cm and 15-30 cm. The

BS treatment also has effect on the production of N_2O and the UREA treatment has an effect on NH_4^+ at a depth 0-15 cm and a slight effect on CO_2 production. These results were found to correspond with a production rate of N_2O that was ≤ 2.1 µg N kg^{-1} day^{-1} with a maximum production in the first 14 days of more than that in the UREA treatment (< 1.5 µg N kg^{-1} day^{-1}). Additionally, the PCA showed that *B. subtilis*, as a regular strain, had a marked effect on the production of N_2O but not on CO_2 production. This strain might be involved in nitrifier denitrification (or aerobic nitrification-denitrification) as was reported by Kim et al. [10] and Yang et al. [11]. Both research groups demonstrated that the *Bacillus* genus is involved in nitrification and denitrification, namely *B. subtilis*, *B. cereus* and *B. licheniformis*, where *B. subtilis* is involved in the nitrification process and *B. licheniformis* is involved in denitrification or aerobic denitrification [10]. Yang et al. reported that the strain of *Bacillus subtilis* A1 is an aerobic heterotrophic nitrifying–denitrifying bacterium, which is able to convert NH_4^+ to N_2 under fully aerobic conditions, while growing either autotrophically or heterotrophically [11]. In our experiment, the environmental condition of the soil was aerobic throughout the experiment, so it could be hypothesized that the *B. subtilis* strain could be involved in the nitrification process or the nitrifier denitrification process (or aerobic denitrification).

4.2. Second experiment under greenhouse conditions

The second experiment was carried out using wastewater sludge as an organic fertilizer. It tested the effect of the sludge or urea on sunflower growth, and the effect of some soil properties on the production of CO_2 and N_2O. The plant characteristics were also evaluated. Wastewater sludge or sewage sludge is generated during wastewater treatment and is an unavoidable by-product. However, the sludge can be seen as an invaluable by-product when it is applied to the soil as stabilized sludge. Also, this waste management is the most economical form of disposal employed to reduce the large amount of sewage sludge. In addition, waste-water sludge is organic matter rich in minerals and is an outstanding source of C and N, *inter alia*. Applying sludge to the soil offers the opportunity of recycling nutrients for use by plants, while at the same time returning C as organic matter to the soil in order to improve agriculture processes.

This experiment was carried out in the same way as the previous experiment (described above). The wastewater sludge was collected from the Reciclagua treatment plant, S.A. de C.V., where wastewater from various industries (including the food industry) and households is treated. The properties of the sewage sludge were pH 8.1, EC 7.9 dS m^{-1}, water content 847 g kg^{-1}, organic C content 288 g kg^{-1}, total N content 41.8 g kg^{-1}, NH_4^+ 13 g N kg^{-1}, NO_2^- 8.3 mg N kg^{-1}, and NO_3^- 122 mg N kg^{-1}. For more details of these, see [12]. Four treatments were established in cylindrical pots, comprising nine soil samples from three sampled plots ($n = 27$). The treatments were: i) unfertilized and unsown soil (CONTROL treatment), ii) unfertilized soil cultivated with sunflowers (SUNFLOWER treatment), iii) cultivated soil fertilized with 0.5 g urea (0.5 g urea × 2 applications, equivalent to 150 kg N ha^{-1}, UREA treatment), and iv) soil cultivated and fertilized with 30 g sludge (SLUDGE treatment). The sludge was added so as to be equivalent to 150 kg N ha^{-1}, assuming that sludge mineralisation was 40% mineral N during the crop cycle. Tap water supplied a total amount of 19 kg mineral-N ha^{-1} through irrigation in all treatments and throughout the experiment. In order to measure the production of gases, the pots were closed airtight approximately every two days for the first 30 days, and their atmospheres were analysed for CO_2 and N_2O at sequential times of 0, 3, 15 and 30 mins. The experiment was replicated twice [12].

The results showed that the CONTROL, SUNFLOWER and UREA treatments were not significantly different with respect to the production rate of CO_2 (1.59, 2.03 and 2.6 mg C kg^{-1} day^{-1}, respectively) (Table 1). The CO_2 production rate of sunflowers cultivated in soil fertilized with sewage sludge (SLUDGE treatment) was 2.96 mg C kg^{-1} day^{-1} and was significantly different compared with the CONTROL treatment. In other words, soil cultivated and fertilized with sewage sludge was equivalent to both the soil cultivated and fertilized with urea, and the unfertilized soil. It should be taken in account that several factors affect CO_2 production in the soil, such as rhizosphere respiration and soil microbial respiration, soil moisture, soil temperature, substrate quantity and quality, vegetation type, and land use and management regimes [13].

Treatment	CO_2 mg C kg^{-1} day^{-1}	N_2O μg N kg^{-1} day^{-1}
CONTROL	1.59 B	0.18 B
SUNFLOWER	2.03 AB	0.27 B
UREA	2.60 AB	0.66 B
SLUDGE	2.96 A	2.50 A
Least significant difference	1.14	0.74
Standard error of estimate ($P < 0.05$)	0.41	0.26

Values with the same letter show no significant difference between treatments ($P < 0.05$).

Table 1. The production rates of CO_2 and N_2O from soil cultivated with *H. annuus* under greenhouse conditions.

On the other hand, the rate of production N_2O was significantly different in cultivated soil fertilized with sludge, 2.5 μg N kg^{-1} day^{-1} compared with the remaining treatments, 0.7, 0.3, and 0.2 μg N kg^{-1} day^{-1} for the UREA, SUNFLOWER, and CONTROL treatments respectively. As previously discussed, wastewater sludge is an organic matter which is rich in easily decomposable material. In addition, it has been demonstrated that microorganisms from the sludge plus those from the soil work together synergistically to accelerate the decomposition of organic matter [6]. The high levels of microbial activity stimulated by the addition of material with high C and N contents could increase the production of both CO_2 and N_2O. According to Kool et al. [9], the loss of N in our experiment might primarily be as a result of the nitrifier nitrification process (from ammonium oxidation) followed by the nitrifier denitrification process. As a result, the wastewater sludge had a NH_4^+ concentration of 13 g N kg^{-1}, an amount of ammonium which might not be oxidized to N_2O so rapidly. The chemical composition of organic or mineral fertilizers — or even residues applied to the soil — is an important factor in regulating the magnitude of N_2O production.

A PCA was performed on all relevant properties of the soil. The first principal component explained about 22% of the observed variation, while the second accounted for 17% of the observed variation. On the related scatter plot, the UREA treatment lies in upper right quadrant and the SLUDGE treatment is in the upper left quadrant. The SUNFLOWER and CONTROL treatments were found in lower left and right quadrants respectively (Figure 1).

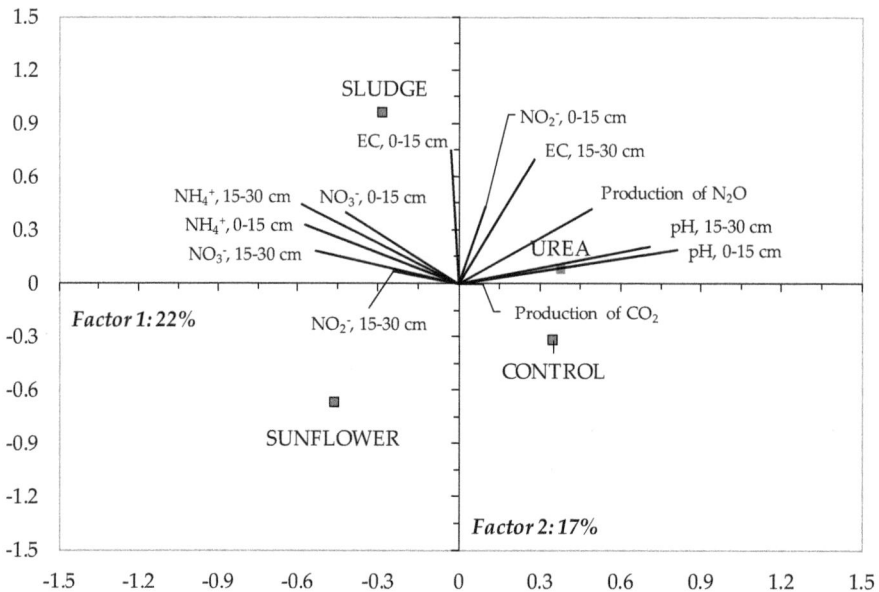

Figure 1. Results of a principal component analysis (PCA) performed on soil properties under greenhouse conditions ($n = 27$).

The UREA treatment seems to have an effect on the pH of soil at both depths (0–15 and 15–30 cm) and also affects the production of CO_2 and N_2O. The PCA also revealed that the SLUDGE treatment affected EC at a depth of 0–15 cm and had a slight effect on both ammonium and nitrate concentrations at both depths.

4.3. Third experiment under greenhouse conditions

In this study, the effect of urea, wastewater sludge and vermicomposting on the production of CO_2 and N_2O was investigated. The Otumba soil (State of Mexico, Mexico), which was characterized as a sandy loam with pH 7.6, EC 1.15 dS m^{-1}, and an organic C content of 7.2 g C kg^{-1}, was used in this study. Wastewater sludge was again collected from Reciclagua S.A. de C.V. (as described above). The vermicompost was prepared with wastewater sludge from Reciclagua and *Eisenia fetida*. The vermicompost was obtained from a mixture of sludge (1,800 g) and manure (800 g) with 70% water content, and was added to 40 individuals of *E. fetida* with the mixture being conditioned over three months. The properties of the vermicompost were pH 7.9, EC 11 dS m^{-1}, organic C content 163 g kg^{-1}, and a total N content of 2 g kg^{-1}. As well, the vermicompost presented < 3 CFU g^{-1} *Salmonella* sp., no *Shigella* sp., and no helminth ova.

Forty-five sub-samples of 3.25 kg soil were prepared, i.e. three soil samples, three replicates, and five treatments were established. The treatments were: i) soil fertilized with 0.07 g urea kg^{-1}, (UREA treatment), ii) soil fertilized with 21.2 g sewage sludge kg^{-1} (H-SLUDGE treatment), iii) soil fertilized with 12.8 g sewage sludge kg^{-1} (L-SLUDGE treatment), iv) soil fertilized with 81.5 g vermicompost kg^{-1} (VERMI treatment), and v) unfertilized soil (CONTROL treatment). The amount of urea was 80 kg N ha^{-1} for the UREA, H-SLUDGE, and VERMI treatments, while the L-SLUDGE treatment was 48 kg N ha^{-1}. The production of CO_2 and N_2O was analysed every two days at 0, 3, 15 and 30 mins, until day 97. The cultivar of common bean used was Negro-8025 (from Universidad Autónoma Chapingo, Texcoco, State of Mexico). Tap water was used for irrigation at a rate of 500 mL every seven days. The experiments were triplicated and ran for 117 days in total.

A large amount of CO_2 was recorded at the end the experiment (from day 45 until day 82). The UREA treatment had no significant effect on CO_2 production when compared with the CONTROL treatment (0.043 mg C kg^{-1} dry soil). Wastewater sludge was found to increase the mean production of CO_2 in the soil at 0.064 mg C kg^{-1} compared with the untreated soil, and the VERMI treatment showed the largest mean CO_2 production at 0.1 mg C kg^{-1}. Urea had no significant effect on CO_2 production when compared with the CONTROL treatment. Occasionally, urea might stimulate the production of CO_2 in N depleted soils. Wastewater sludge increased CO_2 production through the mineralization of organic C and the increasing of microbial activity in the rhizosphere. Similarly, vermicomposting stimulated plant growth and increased root exudates and microbial activity in the rhizosphere. The high levels of CO_2 produced towards the end of the experiments could be related to rapid root growth and the beginning of root decomposition. For example, common bean plants at 49 days after sowing (flowering and nodule senescence begins) showed a decrease in the number of nodules, and their nodule cell walls slowly became thinner and degraded [14] thereby improving the environment for the growth of microorganisms and increasing the production of CO_2 towards the end of the experiments.

The mean production of N_2O was -0.004 µg N kg^{-1} in the CONTROL treatment. Soil fertilized with urea increased the mean production of N_2O to 0.015 µg N kg^{-1}. Wastewater sludge increased the mean production to 0.11 µg N kg^{-1} and 0.58 µg N kg^{-1} in the L-SLUDGE and H-SLUDGE treatments respectively, and the vermicompost mean was 0.32 µg N kg^{-1} dry soil. N_2O production in order of effect size of treatment was: wastewater sludge (H-SLUDGE) > vermicomposting > wastewater sludge (L-SLUDGE) > urea > unfertilized soil. Nitrifier nitrification and nitrifier denitrification were presumably the processes that contributed the most to the production of N_2O under aerobic conditions.

4.4. Fourth experiment under greenhouse conditions

Juárez-Rodríguez et al. applied the sludge derived from anaerobically digested cow manure in the production of biogas (methane-air), to maize (*Zea mays* L.) cultivated in a nutrient-low, alkaline-saline soil with EC 9.4 dS m^{-1} and pH of 9.3. The results showed that the CO_2

production increased 3.5-fold in the soil cultivated with maize and sludge, and increased 3.1-fold after the sludge was added to the soil. The production of CO_2 from soil cultivated with maize showed a 1.6-fold increase compared with the uncultivated and unfertilized soil, 1.5 mg C kg^{-1} day^{-1}. N_2O production was -0.0004 µg N kg^{-1} soil day^{-1} in unfertilized soil, and in soil cultivated with maize was 0.3 µg N kg^{-1} day^{-1}. Soil treated with sludge increased the production of N_2O up to 4.6 µg N kg^{-1} soil day^{-1}. Nevertheless, it was found that cultivated soil produced 2.4 µg N kg^{-1} soil day^{-1}, reducing N_2O production. It was also found that applying the anaerobically digested cow manure stimulated the growth of maize cultivated in an alkaline and saline soil, and the production of CO_2 and N_2O was increased.

4.5. Fifth experiment under greenhouse conditions

In this study, the main aim was to investigate how maize fertilized with wastewater at 120 kg N ha^{-1} affected crop growth, soil properties and the production of carbon dioxide (CO_2), methane (CH_4) and nitrous oxide (N_2O) compared with plants fertilized with urea [16].

The soil was collected from the Mezquital Valley, located near Pachuca in the State of Hidalgo (Mexico). The irrigation water used was slightly alkaline with a pH of 8.4. The experiment was carried out under greenhouse conditions. Soil collected from three sub sites was placed into cylindrical pots. Five treatments were established in order to study the effect of wastewater and urea on the cultivation of maize (*Zea mays* L.). The treatments were: a) SMWW, maize plant plus wastewater; b) SMUREA, maize plant plus urea as fertilizer; c) SUREA, uncultivated soil and urea as fertilizer; d) SWW, uncultivated soil plus wastewater; and e) SCONTROL treatment, soil plus tap water. Soils from the SMWW and SWW treatments were irrigated with 1000 mL of wastewater every 7 days from the first day onwards, making a total of 13 times overall. This means that a total amount of mineral N equivalent to 120 kg N ha^{-1} was added to each maize plant, i.e. the recommended amount of N fertilizer for maize.

The concentration of NH_4^+ was larger in the soil treated with urea and wastewater than in the untreated soil, as the urea was hydrolysed and the wastewater contained high concentrations of NH_4^+. The addition of wastewater to the soil doubled the production of CO_2 and approximately 0.2 g C was produced from the soil due to the decomposition of the wastewater after 70 days. In other words, 34% wastewater C was mineralized. However, urea may only occasionally stimulate CO_2 production when a soil is N depleted. Plants take CO_2 from the atmosphere, but mineralization of root exudates increases the production of CO_2. The production of CO_2 increased towards the end of the period of maize growth. This indicated that the phenological stage of the plants affected CO_2 production. The growth of maize plants was similar under the SMWW (wastewater) and the SMUREA (urea) treatments, even when the release of nutrients was delayed by mineralisation from the organic matter in the wastewater. When wastewater was applied to the soil, the mean production rate of CO_2 increased significantly at 2.4-fold, 1.7 µg C kg^{-1} h^{-1}, compared with the SCONTROL treatment at 0.7 µg C kg^{-1} h^{-1} (Table 2). Meanwhile, cultivating maize increased CO_2 production 3.2-fold, 5.6 µg C kg^{-1} h^{-1}. The SWW, SMWW or SUREA treatments did not show a significant difference in the production of N_2O compared with the SCONTROL (1.5×10^{-3} µg N kg^{-1} h^{-1}). The addition of urea did not affect the CH_4 oxidation rate (0.1×10^{-3} µg C kg^{-1} h^{-1}), nor did the SMUREA treatment (cultivated soil fertilized with urea), but the addition of wastewater to the soil significantly increased CH_4 production to 128.4×10^{-3} µg C kg^{-1} h^{-1}. Soil irrigated with wastewater increased the global warming potential (GWP) up to 2.5-fold compared with the SUREA treatment (soil plus urea), whereas cultivated soil increased the GWP 1.4-fold. Crops irrigated with wastewater might limit the use of N fertilizer and water from aquifers. Nevertheless, the amount of fertilizer applied must be limited due to nitrate (NO_3^-) leaching and the production of CO_2, N_2O and CH_4 – that they could be produced in significant amounts –, and at the same time the salt content of the soil will accumulate, limiting the growth of the crop.

Soils can be either a net sink or a net source of CH_4, depending on several factors such as the moisture level, N level, and the nature of the ecosystem in question. Methane is used up by methanotrophic microorganisms, which are ubiquitous in several soils, and is produced by methanogenic microorganisms in the soil under anaerobic conditions. Agricultural systems are not normally large sources or sinks of CH_4. Only under certain conditions are they sources of CH_4 — after application of manure or other organic materials, or moderate to high levels of irrigation. Our results showed that soil irrigated with wastewater — with or without maize — increased CH_4 production significantly (SMWW and SWW treatments) particularly after irrigation, due to temporary anaerobic conditions.

Treatment	CO_2 (μg C kg^{-1} h^{-1})	N_2O (μg C kg^{-1} h^{-1})	CH_4 (μg C kg^{-1} h^{-1})	GWP [a] (g C kg^{-1} soil)
SMWW	5.6 A[b]	2.8×10^{-3} A	163.6×10^{-3} A	1.97
SMUREA	4.95 A	4.5×10^{-3} A	8.4×10^{-3} C	1.44
SUREA	0.9 C	3.3×10^{-3} A	0.1×10^{-3} C	0.36
SWW	1.7 B	2.5×10^{-3} A	128.4×10^{-3} B	0.90
SCONTROL	0.7 C	1.5×10^{-3} A	1.5×10^{-3} C	0.26

[a] The global warming potential (GWP) of the gases produced was calculated considering CO_2 production equivalent to 310 for N_2O, 21 for CH_4 and 1 for CO_2 (IPCC, 2007) over a 90-day period, minus the C that was stored in the roots per kg soil.

[b] Values with the same letter show no significant difference between treatments ($P < 0.05$).

Table 2. Production of greenhouse gases, CO_2, CH_4 (μg C kg^{-1} soil h^{-1}), and N_2O (μg N kg^{-1} soil h^{-1}) from five treatments: a) soil + plant + wastewater (SMWW), b) soil + plant + urea (SMUREA), c) soil + urea (SUREA), d) soil + wastewater (SWW), and e) soil + water (SCONTROL).

Fertilizing maize with urea or wastewater had a similar effect on plant growth, so wastewater might be useable as a crop fertilizer. The treatments with urea or wastewater had no effect on the pH of soil in this experiment due to the fact that the soil is a vertisol, characterized by a clay type 2:1, with a large capacity for the exchange of protons and consequently, a high buffering capacity. The addition of wastewater increased the production of both CO_2 and CH_4 compared with the soil treated with urea, but did not increase the production of N_2O. The irrigation of crops with wastewater might in the long term be a far more environmentally friendly approach to that of using water from aquifers that take long periods of time to fill, as long as the amount of wastewater applied is restricted to the amount required by the cultivated crop due to possible substantial losses of mineral N through several process such as the production of CO_2, CH_4 and N_2O, and the fact that soil salinization could increase rapidly.

5. Conclusions

The organic fertilizers or treatments (vermicompost, wastewater sludge, anaerobically digested cow manure, and wastewater) might increase the production of greenhouse gases, as do several abiotic and biotic factors involved in microbial activity within the soil. Sludge as a soil fertilizer offers the opportunity for the recycling of plants nutrients and the recovery of C as organic matter and its use in soil to improve agriculture. The high levels of microbial activity stimulated by the addition of this high C and N content material could increase both CO_2 and N_2O production. It should be taken in account that several factors are involved in the production of gases in the soil such as rhizosphere respiration, vegetation type, and soil microbial respiration, as well as abiotic factors such as soil moisture, soil temperature, substrate quantity and quality, and land use and management regimes.

Environmental and economic implications must be considered in order to make well-informed decisions on the management of soil treatments, i.e. how many, how often and what kind of organic fertilizer should be used in order to improve crop production and simultaneously limit soil deterioration and greenhouse gases production.

Acknowledgements

We would like to thank the Instituto Politécnico Nacional and CONACyT for their financial support and the grant-aided support received.

Author details

Fernando López-Valdez [1*], Fabián Fernández-Luqueño [2], Carolina Pérez-Morales [1] and Mariana Miranda-Arámbula[1]

*Address all correspondence to: flopez2072@yahoo.com

1 Agricultural Biotechnology Group, Research Centre for Applied Biotechnology, Instituto Politécnico Nacional, Tlaxcala, Mexico

2 Sustainability of Natural Resources and Energy Program, Cinvestav-Saltillo, Saltillo, C.P. Coahuila, Mexico

References

[1] Tilman D, Balzer C, Hill J, Befort BL. Global food demand and the sustainable intensification of agriculture. P Natl Acad Sci USA.. 2011;108:20260-20264. DOI: 10.1073/pnas.1116437108

[2] Tilman D. Global environmental impacts of agricultural expansion: The need for sustainable and efficient practices. P Natl Acad Sci USA. 1999;96:5995-6000. DOI: 10.1073/pnas.96.11.5995

[3] Gregorich EG, Rochette P, VandenBygaart AJ, Angers DA. Greenhouse gas contributions of agricultural soils and potential mitigation practices in Eastern Canada. Soil Till Res. 2005;83:53-72. DOI: 10.1016/j.still.2005.02.009

[4] López-Valdez F, Fernández-Luqueño F, Luna-Guido ML, Marsch R, Olalde-Portugal V, Dendooven L. Microorganisms in sewage sludge added to an extreme alkaline saline soil affect carbon and nitrogen dynamics. Applied Soil Ecology. 2010;45:225–231. DOI: 10.1016/j.apsoil.2010.04.009

[5] Schimel J, Holland EA.Global gases. In: David M. Sylvia, Jeffry J Fuhrmann, Peter G. Hartel, David A. Zuberer, editors. Principles and Applications of Soil Microbiology. 2nd ed. USA: Pearson Prentice Hall; 2005. p. 491-509.

[6] López-Valdez F, Fernández-Luqueño F, Ceballos-Ramírez JM, Marsch R, Olalde-Portugal V, Dendooven L.A strain of *Bacillus subtilis* stimulates sunflower growth (*Helianthus annuus* L.) temporarily. Scientia Horticulturae. 2011;128:499–505. DOI:10.1016/j.scienta.2011.02.006

[7] Phillips RL, Podrebarac F. Net fluxes of CO_2, but not N_2O or CH_4, are affected following agronomic-scale additions of urea to prairie and arable soils. Soil Biology & Biochemistry. 2009;41:2011–2013. DOI: 10.1016/j.soilbio.2009.06.014

[8] Wrage N, Velthof GL, van Beusichem ML, Oenema, O.Role of nitrifier denitrification in the production of nitrous oxide. Soil Biol. Biochem. 2001;33:1723-1732. DOI: 10.1016/S0038-0717(01)00096-7

[9] Kool DM., Dolfing J, Wrage N, Van Groenigen JW. Nitrifier denitrification as a distinct and significant source of nitrous oxide from soil. Soil Biology & Biochemistry. 2011;43:174-178. DOI: 10.1016/j.soilbio.2010.09.030

[10] Kim JK, Park JK, Cho KS, Nam SW, Park TJ, Bajpai R. Aerobic nitrification–denitrification by heterotrophic Bacillus strains. Bioresource Technology. 2005;96:1897–1906. DOI: 10.1016/j.biortech.2005.01.040

[11] Yang XP, Wang SM, Zhang DW, Zhou LX. Isolation and nitrogen removal characteristics of an aerobic heterotrophic nitrifying–denitrifying bacterium, *Bacillus subtilis* A1. Bioresource Technology. 2011;102:854–862. DOI: 10.1016/j.biortech.2010.09.007

[12] López-Valdez F, Fernández-Luqueño F, Luna-Suárez S, Dendooven L. Greenhouse gas emissions and plant characteristics from soil cultivated with sunflower (*Helianthus annuus* L.) and amended with organic or inorganic fertilizers. Science of the Total Environment. 2011;412-413:257-264. DOI: 10.1016/j.scitotenv.2011.09.064

[13] Ding W, Meng L, Yin Y, Cai Z, Zheng X. CO_2 emission in an intensively cultivated loam as affected by long-term application of organic manure and nitrogen fertilizer. Soil Biology & Biochemistry. 2007;39:669–679. DOI: 10.1016/j.soilbio.2006.09.024

[14] Fernández-Luqueño F, Dendooven L, Munive A, Corlay-Chee L, Serrano-Covarrubias LM, et al. Micro-morphology of common bean (*Phaseolus vulgaris* L.) nodules undergoing senescence. Acta Physiologiae Plantarum. 2008;30(4):545-552. DOI: 10.1007/s11738-008-0153-7

[15] Juárez-Rodríguez J, Fernández-Luqueño F, Conde E, Reyes-Varela V, Cervantes-Santiago F, Botello-Alvarez E, et al. Greenhouse gas emissions from an alkaline saline soil cultivated with maize (*Zea mays* L.) and amended with anaerobically digested cow manure: a greenhouse experiment. Journal of Plant Nutrition. 2012;35(4): 511-523. DOI: 10.1080/01904167.2012.644371

[16] Fernández-Luqueño F, Reyes-Varela V, Cervantes-Santiago F, Gómez-Juárez C, Santillán-Arias A, Dendooven L.Emissions of carbon dioxide, methane and nitrous from soil receiving urban wastewater for maize (*Zea mays* L.) cultivation. Plant Soil. 2010;331:203–215. DOI: 10.1007/s11104-009-0246-0

2

Methane Emissions from Rice Production in the United States — A Review of Controlling Factors and Summary of Research

Alden D. Smartt, Kristofor R. Brye and Richard J. Norman

Additional information is available at the end of the chapter

Abstract

Flooded rice (*Oryza sativa* L.) cultivation has been identified as one of the leading global agricultural sources of anthropogenic methane (CH_4) emissions. Furthermore, it has been estimated that global rice production is responsible for 11% of total anthropogenic CH_4 emissions. Considering that CH_4 has a global warming potential that is approximately 25 times more potent, on a mass basis, than carbon dioxide (CO_2) and rice production is globally extensive and concentrated in several mid-southern and southern states and California, the purpose of this review is two-fold: (i) discuss the factors known to control CH_4 production in the soil and transport to the atmosphere from rice cultivation and (ii) summarize the historic and recent research conducted on CH_4 emissions from rice production in the temperate United States. Though some knowledge has been gained, there is much more that still needs to be learned and understood regarding CH_4 emissions from rice production in the United States, its contribution to climate change, and potential mitigation strategies. Extending the current knowledge base surrounding CH_4 emissions from rice cultivation will help regulatory bodies, such as the Environmental Protection Agency, refine greenhouse gas emissions factors to combat the potential negative effects of climate change.

Keywords: Methane, emissions, rice production, agriculture, soil texture

1. Introduction

Methane (CH_4) is a known and potent greenhouse gas that is produced by anaerobic *Archaea* under anoxic conditions. Agricultural activities have been recognized as contributing an estimated 50% to global anthropogenic CH_4 emissions [1], while an estimated 31% of anthropogenic CH_4 emissions have been attributed to agricultural activities in the United States (US)

[2]. Due to the anaerobic conditions that form in saturated soils, which is a prerequisite for CH_4 production, flooded rice (*Oryza sativa* L.) cultivation has been specifically identified as one of the leading global agricultural sources of anthropogenic CH_4 emissions, accounting for approximately 22% of the total global agriculturally related CH_4 emissions [3]. Furthermore, it has been estimated that global rice production is responsible for 11% of total anthropogenic CH_4 emissions [1,3].

While numerous factors have been determined to impact CH_4 emissions from rice cultivation, due to a general lack of field data, the United States Environmental Protection Agency (USEPA) currently uses a single emissions factor for all non-California-grown, primary rice crops [4]. Therefore, the purpose of this review is two-fold: (i) discuss the factors known to control CH_4 production in the soil and transport to the atmosphere from rice cultivation and (ii) summarize the historic and recent research conducted on CH_4 emissions from rice production in the temperate United States.

2. The greenhouse effect

The greenhouse effect is a mechanism by which certain gases such as carbon dioxide (CO_2), CH_4, nitrous oxide (N_2O), and water (H_2O) vapor absorb and release infrared radiation, interfering with the ability of solar radiation to leave Earth's atmosphere. The absorption of thermal radiation by H_2O and CO_2 was discovered through laboratory experiments in 1859 [5]. However, other gases including CH_4 and N_2O were not recognized as greenhouse gases until the 1970s [6].

Global warming potential (GWP) is a metric that allows the warming impact of various greenhouse gases to be quantitatively compared on the same scale. The assignment of GWP values to gases requires knowledge of the contribution to global warming of gas emissions over time based on the amount of radiation per mass that the gas can absorb and emit as well as the atmospheric lifetime of the gas. Global warming potentials are assigned relative to that of CO_2, thus the 100-yr GWP of CO_2, CH_4, and N_2O are 1, 25, and 298, respectively [7]. For example, 1 kg of CH_4 released to the atmosphere is equivalent to 25 kg of CO_2 being released. Global warming potentials allow greenhouse gas emissions to be reported as CO_2 equivalents in order to compare warming effects of various gases on a single scale.

The current climate change problem is not a result of the greenhouse effect itself, but rather from an increasing greenhouse effect resulting from anthropogenic activities that have increased atmospheric concentrations of greenhouse gases. Prior to 1750, the atmospheric CO_2 mixing ratio was about 280 parts per million (ppm) [8]. Since the beginning of the industrial era, atmospheric CO_2 has risen drastically to 379 ppm in 2005 [7] and 395 ppm as of April 2013 [9]. Between 1750 and 2005, atmospheric CH_4 increased from about 700 parts per billion (ppb) to 1,774 ppb [7]. Nitrous oxide was more variable ranging from 180 to 260 ppb prior to 1750, but has similarly increased to a mixing ratio of 319 ppb in 2005 [7]. While atmospheric N_2O and CO_2 concentrations have increased steadily over the past several decades, the growth rate (i.e., concentration increase) of atmospheric CH_4 seems to be declining. The growth rate of atmospheric CH_4 has decreased from highs of about 1% per year in the 1970s and 1980s to nearly zero between 1999 and 2005. However, the decreasing growth rate is poorly understood [7].

3. Greenhouse gas emissions

Globally, CO_2 accounted for about 76% of greenhouse gas emissions in 2004, with around 75% of CO_2 emissions resulting from fossil fuel use and much of the remainder from deforestation and biomass decomposition [10]. Methane and N_2O accounted for 14 and 8%, respectively, of the estimated global greenhouse gas emissions in 2004. Major CH_4 sources include agricultural activities, waste management, and energy use, while N_2O emissions are primarily a result of agricultural activities, such as fertilizer use and soil management [10]. In the US in 2013, an estimated 82% of the total greenhouse gas emissions were CO_2, 10% were CH_4, and 5% were N_2O [2]. Major sources of greenhouse gas emissions are generally the same in the US as the global sources mentioned above. The major global sectors responsible for greenhouse gas emissions are energy supply (26%), industry (19%), forestry (17%), agriculture (14%), and transportation (13%) [10]. In comparison, the major US sectors responsible for greenhouse gas emissions are energy supply (31%), transportation (27%), industry (21%), commercial and residential (12%), and agriculture (9%) [2].

Although agricultural activities do not dominate total greenhouse gas emissions, agriculture contributes an estimated 50 and 60% of global anthropogenic emissions of CH_4 and N_2O, respectively [1]. Agriculture in the US is responsible for an estimated 36% of anthropogenic CH_4 emissions and 79% of anthropogenic N_2O emissions [2]. Enteric fermentation, rice cultivation, and manure management contribute an estimated 64, 22, and 8%, respectively, to global anthropogenic agricultural CH_4 emissions, while agricultural N_2O emissions are dominated by agricultural soil management (80%) [3]. In comparison, enteric fermentation, rice cultivation, and manure management contribute to 70, 4, and 26% of US anthropogenic agricultural CH_4 emissions [2]. Although rice cultivation makes up a small portion of CH_4 emissions in the US, globally rice cultivation accounts for approximately 11% of total anthropogenic CH_4 emissions.

Methane emissions from US rice cultivation were estimated to be 8.3 Tg CO_2 equivalents in 2013, a reduction from 9.3 Tg CO_2 equivalents in 2012 due to a decline in rice production area [2]. Arkansas was responsible for 36% of the estimated CH_4 emissions from rice cultivation, although Arkansas accounted for 43% of the total US rice production in 2013. Louisiana was the next leading contributor to CH_4 emissions accounting for 27% of 2013 emissions, while harvesting 16% of 2013 production [2,11]. Louisiana and Texas CH_4 emissions are large relative to their production areas due to extensive ratoon cropping in 2013, which occurred on an estimated 38 and 68%, respectively, of the production area in those states [2]. A ratoon crop is a second crop that is managed and produced after the first or primary crop is harvested. California, Mississippi, and Missouri, none of which reported any ratoon cropping, contributed 14, 3.6, and 4.5%, respectively, to the estimated 2013 CH_4 emissions from US rice cultivation [2].

The USEPA periodically publishes CH_4 emissions factors based on research data. Separate emission factors of 178 kg CH_4-C ha^{-1} season^{-1} and 585 kg CH_4-C ha^{-1} season^{-1} were used in the inventory estimates for non-California-grown, primary rice cropping and ratooned cropping areas, respectively, as is consistent with the Intergovernmental Panel on Climate Change [3], which recommends calculating separate emissions factors for as many different factors and cultural practices as is possible. Emissions factors for California rice production are 200 and 100 kg CH_4-C ha^{-1} season^{-1} for winter-flooded and non-winter-flooded rice, respectively [2]. While it is known that factors such as water management, soil properties, rice cultivar, fertilizer management, and residue management have strong impacts on CH_4 emissions from rice cultivation, data available from US studies limit the further disaggregation of these factors [2]. The non-California-grown, primary crop emissions factor is based on US studies with emissions ranging from 46 to 375 kg CH_4-C ha^{-1} season^{-1} [13–20] and the ratoon crop factor is based on studies conducted in Louisiana with emissions ranging from 361 to 1118 kg CH_4-C ha^{-1} season^{-1} [21,22]. The California-specific emissions factors include studies with emissions ranging from 47 to 166 kg CH_4-C ha^{-1} season^{-1} for the non-winter-flooded and from 98 to 277 kg CH_4-C ha^{-1} season^{-1} for the winter-flooded rice [23,24].

4. Rice production

Rice is a semi-aquatic, cereal grain that makes up about 21% of total global grain production [25]. The importance of rice is further exemplified by the fact that rice is a staple food crop for about half of the global population, with direct human consumption accounting for 85% of rice production compared to 72% of wheat (*Triticum aestivum* L.) and 19% of maize (*Zea mays* L.) production [26,27]. In Southeast Asia, 60% of human food intake is provided by rice as well as 35% of food intake in both East Asia and South Asia [26]. Rice has the ability to support more people per unit of land area than wheat or maize because rice produces, on an average yield basis, more food energy and protein per hectare than wheat or maize [28]. Therefore, any potential negative environmental consequences associated with rice production have to be taken seriously.

4.1. Rice production extent

Common rice (*Oryza sativa*) is commercially produced in 112 countries worldwide, spanning latitudes from 53°N along the Amur River at the China–Russia border to 35°S in central Argentina [26]. In 2012, more than 158 million ha globally were planted to rice, with average yields of 4.4 Mg ha^{-1} for a total global production of 470 Tg of rice. Comparatively, nearly 216 million ha were planted in wheat in 2012, with average yields of 3.0 Mg ha^{-1} for a total of 656 Tg of global wheat production. More than 174 million ha were planted in maize in 2012, with an average yield of 4.9 Mg ha^{-1} and a total global production of 857 Tg of maize [25]. Global rice production peaked in 1994 at 534 Tg of rice, with Asia being responsible for 90% of that production [29]. The majority of global rice production occurs in east, south, and southeast Asia, which together accounted for 90% of global production in 2012. Substantial production also occurs in South America (Brazil and Peru), Sub-Saharan Africa (Nigeria and Madagascar), Europe (Italy and Spain), Egypt, and the US [25].

China and India currently dominate global rice production accounting for 30 and 22%, respectively, of the total global production in 2012. The third-, fourth-, and fifth-ranked global producers in 2012 were Indonesia (8%), Bangladesh (7%), and Vietnam (6%). The remaining top 10 producers, in order, were Thailand, the Philippines, Burma, Brazil, and Japan, followed by the eleventh-ranked US, which accounted for 1.3% of global production [25]. The US, however, plays a larger role in global exports contributing 9% of 2012 global exports and ranking fifth after Thailand (21%), India (20%), Vietnam (20%), and Pakistan (10%). Global exports in 2012 were estimated to be 8% of total production, while the US exported 55% of 2012 production [30]. Global rice yields in 2012 were estimated to be 4.4 Mg ha^{-1} compared to 8.3 Mg ha^{-1} in the US, which was second only to Egypt (8.8 Mg ha^{-1}) among the major rice-growing countries. The two top rice-producing countries, China and India, had estimated yields of 6.7 and 3.6 Mg ha^{-1}, respectively [25].

Nearly 1.1 million ha of rice were planted in the US in 2012, yielding an average of 8.3 Mg ha^{-1} for a total production of 9.0 Tg of rice prior to milling, compared to 23 million ha planted with an average yield of 3.1 Mg ha^{-1} for a total of 62 Tg of wheat production, and over 39 million ha of planted maize with average yields of 7.7 Mg ha^{-1} for a total production of 274 Tg [11]. The four major regions that produce rice in the US are the Arkansas Grand Prairie, the Mississippi Delta, which is made up of portions of Arkansas, Missouri, Mississippi, and Louisiana, the Gulf Coast (Texas and southwest Louisiana), and California's Sacramento Valley. Most US states produce primarily long-grain cultivars, while much of the medium-grain rice and nearly all of the short-grain rice is produced in California [11]. Although Oklahoma and Florida are often included as rice-producing states, the six previously mentioned states have made up essentially all of US production in recent years [11]. Arkansas is the leading state in both area of cultivation and total production, contributing 48% of total US rice production in 2012, followed by 23% of production by California and 13% of production by Louisiana [11]. Arkansas rice production takes place in the eastern portion of the state with the top five rice-producing counties in 2012 being Poinsett, Lawrence, Arkansas, Greene, and Cross, which made up 35% of the state's production area [31].

4.2. Global rice production practices

Rice production practices vary globally based on economic, cultural, and climatic factors, each of which show temporal and spatial variability throughout the rice-growing countries. A simple classification or characterization of rice production systems is nearly impossible on a global scale due to the variability of factors that influence production. Classifications of rice production techniques are commonly based upon flood presence (e.g., upland or lowland), water source (e.g., irrigated or rainfed), and stand establishment technique (e.g., transplanting, direct-seeding, or water-seeding) with many combinations and variations of these techniques occurring throughout the globe [32]. In one of the most recent classification attempts, Chang [33] classified global rice production into five major agroecosystems: (i) irrigated wetland, which made up 53% of global rice production area and had the greatest yield potential at 3 to

5 Mg ha^{-1}, (ii) rainfed wetland, making up 26% of global area and yielding 2 to 4 Mg ha^{-1}, (iii) flood-prone or tidal swamps, which made up an insignificant area, (iv) deep water (1–5 m), making up 8% of global area, and (v) dryland, which made up an estimated 13% of global production area with average yield potentials of 1 to 1.5 Mg ha^{-1}.

While a small portion of rice is produced under upland conditions, the majority of rice production requires substantial quantities of water in order to maintain a flood on the semi-aquatic crop. In much of the tropical rice-growing area, particularly south and southeast Asia, rainfed rice is the main production system, where most of the production comes from wet-season harvests and the cropping season is determined by rainfall patterns [32]. In temperate production areas, rice production must coincide with suitable temperatures for the crop which, coupled with inadequate rainfall, requires that temperate rice be almost entirely irrigated in order to maintain a flood for the duration of the growing season [32]. The utilization of irrigation in temperate areas allows greater control of environmental factors, which ultimately tends to increase yields, while rainfed systems may suffer from droughts and floods that may substantially damage crops and reduce yields [32].

Direct-seeding and transplanting are common establishment techniques in both irrigated- and rainfed-wetland systems, while direct-seeding is the major practice in dryland and deep-water agroecosystems [33]. While transplanting does occur in irrigated- and direct seeding occurs in rainfed-wetland systems, it is more common for irrigated systems to utilize direct-seeding and for rainfed systems to use transplanting techniques [32]. Transplanting systems involve raising seedlings in a nursery seedbed area at the beginning of the season and transplanting into puddled paddy soils early in the vegetative growth stage. Transplanting is the major establishment system for rainfed rice in tropical Asia, with the majority of production in northeast India, Bangladesh, and Thailand relying upon transplanting techniques [32]. Direct-seeding by grain-drilling or broadcasting pre-germinated seeds onto puddled soil is practiced in parts of India, Sri Lanka, Bangladesh, and the Philippines, while drill-seeding into dry soil is the most common practice in the US and other mechanized regions such as Australia [32]. Rice seed may be broadcast onto dry or moist soil by airplane followed by harrowing to cover seeds, but this establishment method requires more seed and stand establishment is often poorer than with drill-seeding [32]. Water-seeding is an establishment technique that originated and is practiced in parts of Asia, where pre-germinated seeds are broadcasted from an airplane into already flooded paddies or fields [32]. The rice-production system, and associated specific production practices, can significantly affect CH$_4$ production and emissions.

4.3. Rice production practices in the US

Rice production under mechanized US systems requires high temperatures, nearly level land, plentiful water, and soils that inhibit percolation of floodwater, so production is limited to Arkansas, Louisiana, Mississippi, Missouri, Texas, California, and Florida [34]. All US rice is produced using high-input, mechanized production practices, but practices vary somewhat from region to region based on differences in climate, soils, weed proliferation, and other factors that influence production. Essentially, all US rice is irrigated and sources of irrigation water include shallow or deep groundwater, runoff reservoirs, rivers, bayous, and lakes [34].

It is estimated that between 1000 and 2500 m^3 ha^{-1} of water are required to produce a rice crop in the southern US and generally less than one third of that requirement is met by rainfall [35]. Levees, which separate fields into bays, or paddies, and control flood depth (i.e., by use of gates or spills), are commonly constructed on contours that were surveyed on 3 to 6 cm vertical intervals. This creates winding, contour-shaped levees in fields that are not precision-leveled, whereas precision leveling to a uniform grade of 0.2% or less allows the construction of uniformly spaced, straight levees and may reduce the number of levees required [34].

The two stand establishment techniques utilized in the US are dry-seeding and water-seeding. Dry-seeding techniques, particularly drill-seeding, are predominant in most of the US, while water-seeding techniques are used extensively in California and to a small degree in southwest Louisiana and other regions as a weed control method [34]. A continuously flooded, water-seeding technique is used in California, where pre-germinated seeds are broadcast by airplane into flooded fields and the seedlings grow through a standing flood, while a pinpoint-flood, water-seeding technique is used in Louisiana, where seeds are broadcast into a flooded field that is drained within a few days and then permanently flooded after drying for 3 to 5 days [34,36]. In dry-seeded systems, seed is most often drilled into a well-pulverized, firm, and weed-free seedbed in 15- to 25-cm rows to a depth of 2.5 cm or less. When rice is following a high-residue crop, such as rice, maize, or wheat, it is necessary to till the land in the fall or early spring so that decomposition of the residue does not immobilize nutrients after the subsequent rice crop is planted, whereas rice following soybean (*Glycine max* L.), a crop that produces relatively little residue, may not require as much preparation because crop residues are not as abundant or as persistent compared to that of rice or maize [34,37].

Water management at and shortly after planting varies across US systems, but a permanent flood is established in all systems usually by the four- to five-leaf vegetative growth stage/ beginning tillering (V4-5) [38]. Flush irrigation is used as necessary to promote germination and seedling growth in dry-seeded rice systems prior to establishment of a permanent flood, which typically occurs three to four weeks after emergence (i.e., the V4 to V5 growth stage). Drainage during the season is typically avoided except if a nutrient deficiency, such as zinc, is detected, to aerate the soil in order to treat or prevent disorders, such as straighthead and hydrogen sulfide toxicity, or to apply pesticides. Fields are drained prior to harvest in order to dry the soil enough for operation of harvest equipment [34]. Fields are flooded again within five to seven days after primary-crop harvest in ratoon cropping systems, which are common in southwest Louisiana, Texas, and Florida, and the flood is again maintained until harvest of the ratoon crop [34].

Crop rotations are important in rice, especially where weedy/red rice is problematic and difficult to control during rice cropping seasons. In order to suppress weedy rice, nearly all rice in Louisiana is grown either in a 1:1 rotation with soybean or a 1:1:1 rotation where crawfish (*Procambarus clarkia*) are double-cropped following rice, with soybean produced the following season [34]. In 2012, greater than 70% of Arkansas rice was produced in rotation with soybean, with most of the remaining production in a rice–rice rotation [39]. In California, approximately 70% of rice is produced in a rice–fallow or rice–rice rotation [40].

4.4. Arkansas rice production practices

Arkansas is the leading rice-producing state, accounting for 40 to 50% of total annual production in the US [11]. Rice production in Arkansas began in 1902 when 0.4 ha were planted in Lonoke County. Production increased over time until 1955 when government quotas limited production to 202,350 ha. The limitation was lifted in 1974 and production increased again, peaking in 1981 at 623,240 ha, again in 1999 with 667,755 harvested hectares, and finally in 2010 with 724,413 ha [31]. In 2012, 518,016 ha rice were harvested in Arkansas [11]. Rice production in Arkansas is highly mechanized with a heavy dependence upon synthetic fertilizers, chemical pest control, and machinery. Planting of rice in Arkansas generally begins the last week of March and extends into early June with floods typically being established by the end of May or early June. Harvesting operations usually begin in mid-August and peak in early- to mid-September [31].

Arkansas rice is produced on a wide variety of soils ranging from sandy to clay soils with the differing textural classes generally requiring different management, especially with regards to tillage practices and nutrient management [39, 41]. Production on sands and sandy loams is minor and has been decreasing from 3.1 and 5.2% of Arkansas area, respectively, in 2007 to 0.7 and 3.7%, respectively, in 2012. Arkansas production on clay and clay-loam soils, however, has increased from under 40 to 48% between 2007 and 2009 but declined to 43% in 2012. Production on silt-loam soils has remained fairly steady at 52% in 2007 and 53% in 2012 [39,42].

Dry-seeding techniques have always dominated in Arkansas. Water-seeding has varied between 2 and 8% of the production area between 2007 and 2012, with an estimated 5% of the 2012 Arkansas rice area being water-seeded [39,42]. Approximately 80% of 2012 Arkansas rice area was drill-seeded, compared to approximately 20% being broadcast-seeded [39]. Conventional tillage accounted for over half of Arkansas planted-rice area, while stale-seedbed (i.e., tillage and floating, or leveling the field, in the fall or winter) and no-tillage accounted for 35 and 10% of the planted-rice area, respectively, in 2012 [39]. Stale-seedbed and no-tillage are oftentimes utilized on clay soils where conventional tillage produces a cloddy seedbed with poor seed-to-soil contact [41].

While pinpoint, water-seeding techniques do occur in Arkansas, over 90% of the Arkansas rice production area utilizes a delayed-flood system, where the permanent flood is not established until the four- to five-leaf growth stage, which generally occurs approximately three to four weeks after emergence [39]. Fields are drained two to three weeks prior to harvest and most fields remain unflooded until the subsequent rice crop is produced, while nearly 20% of Arkansas rice area is winter-flooded [34,39]. Over 75% of Arkansas rice is irrigated by groundwater with 10 and 13% of the rice area utilizing water stored in reservoirs and from streams/rivers, respectively [39].

The two methods of nitrogen (N) fertilization in Arkansas are (i) the standard two-way split system, where 65 to 75% of the total N is applied pre-flood with the remainder applied at mid-season in one or two applications between beginning internode elongation and half-inch internode elongation [i.e., reproductive stage 0 (R0) to 1 (R1)], and (ii) the single optimum pre-flood system, where a single N application is made immediately prior to flooding. Nitrogen fertilizer rate recommendations have previously been based only on cultivar, soil texture, and previous crop. Implementation of the new N-Soil Test for Rice (N-STaR) enables recommendations to be adapted to the soil's ability to supply N to the rice crop on a field-by-field basis, reducing the likelihood of over- and under-fertilization of N [43]. Ammonium-N sources, such as urea and ammonium sulfate, are used in order to prevent N loss through denitrification that occurs with nitrate-containing fertilizers. Phosphorus and potassium are incorporated prior to planting as recommended by routine soil tests [43]. Organic amendments are uncommon, although poultry litter is utilized to a small degree, especially in precision-leveled fields.

5. Flooded soils

The saturated soils that occur during wetland, or lowland, rice cultivation give rise to a set of physical, chemical, and biological properties that are quite different from upland soils. Rice is the only major row crop produced under flooded-soil conditions and the absence of air-filled pores along with reduced soil–atmosphere interactions result in an almost entirely different set of processes than those occurring in upland cropping systems.

5.1. Physical characteristics of flooded soils

The major physical difference between saturated and unsaturated soils involves the availability and rates of movement for gases and solutes. Under aerated conditions, the soil atmosphere contains essentially the same gases as the atmosphere although the proportions of oxygen (O_2) and CO_2 differ from the atmosphere due to soil respiration [44]. Carbon dioxide diffuses into the atmosphere from the soil due to production during respiration and O_2 diffuses into the soil as it is consumed during respiration. The saturation and ponding of flooded soils greatly reduce gas transport between the soil and atmosphere compared to aerated soils and plant-mediated transport of gases by diffusion is often the main exchange mechanism between the soil and atmosphere in saturated or flooded systems [45]. As a flooded soil dries, gases trapped in the soil may escape due to increases in diffusion and convective flow rates that occur as water escapes soil pores.

While solute movement by diffusion may be greater in saturated soils due to an increase in water-filled pore space, diffusion of gases through water is roughly three to four orders of magnitude slower than diffusion of gases through air [46,47]. Both diffusive and convective flow of gases and solutes are related to pore connectivity and tortuosity, so it is expected that movement of gases and solutes are slower in fine-textured soils, such as clays and clay loams, than in coarser-textured soils, such as silt loams and sands, which generally have larger, more connected pores [47]. Convective flow of gases in saturated soils can occur as dissolved gases move with moving soil water, which is dependent largely upon soil texture and structure, and as ebullition, which is where gases escape as bubbles through ponded water [47]. Generally, diffusion dominates gas transport in fine-textured soils, such as clay loams and clays, and diffusion rates typically decrease as particle size decreases, which is due to differences in size, orientation, and shape of soil pore spaces [45,48]. Soil texture also affects the amount of time it takes for a soil to become saturated with infiltration rates in clayey soils estimated to be 1 to 5 mm hr^{-1} compared to 10 to 20 mm hr^{-1} in soils such as silt loams [47]. The amount of time a soil takes to become saturated has an effect on chemical and biological processes that develop as the system becomes anaerobic.

5.2. Soil redox potential

Isolation of flooded soils from the atmosphere and depletion of soil O_2 induces biological and chemical reactions that create anaerobic and reducing conditions rather than the aerobic and oxidized conditions that generally occur in upland soils. Organic matter decomposition slows under anaerobic conditions, but as organic matter is oxidized, transformations such as denitrification and manganese (Mn) and iron (Fe) reduction occur as well as production of gases such as hydrogen sulfide (H_2S) and CH_4. Soil reduction/oxidation (redox) reactions are coupled half-reactions where the oxidation of organic matter, which provides electrons, is coupled with the reduction of elements or compounds that act as electron acceptors [49]. Oxygen is the major electron acceptor under aerobic conditions, but as O_2 is depleted, the sequence of electron acceptors shifts to NO_3^-, MnO_2, $Fe(OH)_3$, SO_4^{2-}, and CO_2, which are theoretically reduced in that order based on thermodynamic favorability [44,50]. The reduced forms of the previously mentioned terminal electron acceptors are H_2O, N_2, Mn^{2+}, Fe^{2+}, H_2S, and CH_4, respectively. Soil redox reactions in a controlled laboratory environment may follow the theoretical sequence, but environmental conditions in the field result in spatial variability of oxidizable organic compounds, electron acceptors, and microorganisms that cause substantial overlap of the terminal electron acceptor sequence [44,49].

Soil redox potential (Eh) is a measure of the electrical potential status of a system that results from the tendency of substances in the system to donate or acquire electrons [51]. Soil redox potential is measured in millivolts (mV) using a platinum electrode along with a mercury chloride (HgCl) or silver chloride (AgCl) reference electrode, both connected to a voltmeter [49]. Combination platinum electrodes are also available that can continuously monitor soil Eh when connected to a logger box. When using AgCl electrodes, a correction factor of approximately +200 mV is added to field-measured voltages in order to adjust measurements to the standard hydrogen electrode [52]. In well-aerated soils, soil Eh may be as great as +700 mV, but Eh values near –300 mV may be observed in saturated organic-matter-rich soils [51]. As a system shifts from aerobic to anaerobic and soil redox potential declines, atmospheric O_2 is reduced first at +380 to +320 mV, followed by NO_3^- (+280 to +220 mV), MnO_2 (+220 to +180 mV), $Fe(OH)_3$ (+110 to +80 mV), SO_4^{2-} (–140 to –170 mV), and CO_2 (–200 to –280 mV), based on measurements by Patrick and Jugsujinda [53].

6. Methane emissions from rice

Methane emissions from any ecosystem, particularly a rice agroecosystem (Figure 1), are governed by the magnitude and balance of microbial CH_4 production (methanogenesis) and oxidation (methanotrophy), which occur by separate microbial communities. The two groups of microorganisms are adapted to different environmental conditions, and, as a result, are affected differently based on the structure and conditions of an ecosystem, which results in variability of CH_4 production and oxidation potentials across time and space [54]. With low CH_4 production rates or long diffusion pathways, it seems that the majority of the CH_4 produced is oxidized. Conversely, in cases where CH_4 production rates are high or diffusion paths are short, less CH_4 is oxidized and a greater portion reaches the atmosphere [54] (Figure 1).

Figure 1. Chamber-based measurements of methane emissions from small plots at the Rice Research and Extension Center near Stuttgart, AR (top), and at the Northeast Research and Extension Center at Keiser, AR (bottom). Photographs taken by K. Brye.

6.1. Methane production and oxidation

Methane production occurs toward the end of a complex anaerobic decomposition process in which soil organic matter (SOM) is degraded to acetate, hydrogen gas (H_2), and CO_2 by a community of various fermenting microorganisms, which are mostly bacteria. Methanogenic *Archaea* are then able to split acetate into CH_4 and CO_2 (i.e., acetoclastic methanogenesis) or utilize H_2 and CO_2 to produce CH_4 (i.e., hydrogenotrophic methanogenesis) [55,56]. Methanogens encompass a large group of strictly anaerobic, obligate *Archaea*, which is currently composed of three classes, six orders, 12 families, and 35 genera [56]. Rice Cluster I is a specific group of methanogens identified by Grosskopf et al. [57] that contains enzymes in order to detoxify highly reactive O_2 species, allowing the methanogens to survive in aerated soils or oxygenated rhizospheres, and occurs preferentially in environments that undergo transient aerobic conditions, such as in rice fields [55,58]. Rice Cluster I has been detected in almost all rice field soils tested [59,60] and occurs in great abundance in rice soils and on rice roots, representing up to 50% of total methanogens in rice fields [61]. Rice Cluster I has been identified as occupying a niche on rice roots by producing CH_4 from photosynthates released as root exudates [55,62]. Recent research has confirmed that methanogens are ubiquitous in aerobic soils and have the ability to produce CH_4 as soon as anoxic conditions form and substrate is available [56]. Conrad [63] reported that methanogens isolated from the soil of rice fields were not killed but only inhibited by high redox potentials or O_2 exposure, allowing them to survive drainage and maintain their population size throughout the year in a state of low activity.

Most methanogens are mesophiles and neutrophiles, with optimal growth occurring between 30 and 40°C and between a pH of 6 and 8 [54]. Methanogens are highly sensitive to variations in temperature and pH and CH_4 production is greatly reduced when soil temperatures are low or in acidic or alkaline soils [56]. Within the optimal temperature range, which is generally the case during the rice growing season, temperature has a positive effect on methanogenesis, causing an increase in CH_4 production as temperature increases [54,56].

Methane oxidation is achieved by a group of aerobic *Proteobacteria* known as methanotrophs, which only utilize CH_4 or methanol as a source of C and energy and are currently classified into two phyla, three orders, four families, 21 genera, and 56 species [56]. One group, known as low-affinity methanotrophs, is capable of oxidizing high CH_4 concentrations (>100 ppm) and exists at oxic–anoxic interfaces, where the methanotrophs consume CH_4 produced in anoxic environments [56]. Another group, known as high-affinity methanotrophs, exists in upland soils and possesses the ability to oxidize CH_4 at low atmospheric levels (<2 ppm) [64]. Unlike methanogenesis, methanotrophy is not impacted greatly by temperature, although CH_4 oxidation is decreased below 10°C and above 40°C, or pH, as similar CH_4 oxidation has been observed in soils with pH values ranging from 3.5 to 8 [56]. Due to the differing effect of temperature on methanogenesis and methanotrophy, CH_4 production increases as soil temperatures increase, while CH_4 oxidation changes little, resulting in a general increase in CH_4 emissions as soil temperature increases throughout the rice growing season. This effect has been confirmed in a laboratory incubation of anaerobic soils at various temperatures between 5 and 25°C [65].

6.2. Substrate for methane production

Available SOM stimulates CH_4 production due to enhanced fermentative production of acetate and H_2/CO_2 and, in principle, CH_4 production could be expected to be proportional to organic C inputs, but the reduction of nitrate (NO_3^-), iron (Fe), manganese (Mn), and sulfate (SO_4^{2-}) all precede methanogenesis and reduce the amount of available C for CH_4 production [54]. Methane production may be stimulated by root exudates [66–68] or the application of animal manures [69], green manures [70–73], or rice straw [67,70,73–75], while the application of composted organic C sources does not greatly increase CH_4 production [73,75,76]. This indicates that the amount of available organic C (OC) is more important in determining CH_4 production than total OC (TOC), as composted residue contains lower amounts of degradable C, on a mass basis, compared to fresh residues [77]. Yagi and Minami [73] and Wang et al. [78] confirmed a positive correlation between CH_4 production and readily mineralizable C, while studies have indicated no clear relationship between soil TOC and CH_4 production [68,79–81]. Research conducted by Denier van der Gon and Neue [76] determined that increasing fresh OM inputs would result in increases in CH_4 production up to a point where another factor becomes limiting; however, fresh green manure inputs up to 20 Mg ha^{-1} still indicated OC limitations. In most rice production situations, organic residue inputs are below 20 Mg ha^{-1} and will generally exhibit an increase in CH_4 emissions as organic inputs increase.

Using ^{13}C-labeled rice straw incorporated at 6 Mg ha^{-1}, Watanabe et al. [82] determined that 42% of season-long CH_4 emissions originated from rice straw C, 37 to 40% from the rice plant, and 18 to 21% from SOM. The contribution of SOM to CH_4 production was fairly consistent over the growing season, while the contribution from rice straw decreased from nearly 90% at 14 days after transplanting to only 11 to 16% during heading and grain fill. In contrast, the contribution of living rice plants to CH_4 production increased over time and amounted to 65 to 70% during heading and grain fill [82]. Chidthaisong and Watanabe [83] also observed that the contribution of rice straw to CH_4 production was greatest at 20 to 40 days after flooding, while plant-derived C became increasingly more influential as the season progressed. The link between root exudates and CH_4 production has been observed directly by Aulakh et al. [84], who showed a positive correlation between TOC in root exudates and CH_4 production. Several others have observed an inverse relationship between grain yield and CH_4 production [19,85], indicating that lower grain yields are accompanied by greater CH_4 production as a result of greater root exudation, which was confirmed by Aulakh et al. [66]. Using ^{13}C-labeled CO_2, it was observed that photosynthates were a major source of CH_4 and accounted for 4 to 52% of CH_4 under field conditions [86,87].

6.3. Duration and timing of methane production

Methane production occurs for some period of time following a period of prolonged saturated conditions and continues until the C substrate becomes limiting or environmental conditions limit methanogenesis (i.e., the soil becomes too cold, hot, or aerated). In flooded soils, the rate of reduction processes is determined by the composition and texture of a soil as well as the content of inorganic electron acceptors [i.e., NO_3^-, MnO_2, $Fe(OH)_3$, SO_4^{2-}] and available C, so the amount of time between flooding a soil and the onset of methanogenesis can vary from several days to several weeks [88]. From the onset of methanogenesis, CH_4 emissions from rice systems generally increase over time as the soil becomes more reduced and usually shows one or more of three general peak flux trends. Early season peak fluxes are generally attributed to decomposition of freshly incorporated residues and generally occur within 20 to 40 days after flooding [83,89] and late-season peaks are thought to result from decomposition following senescence of rice roots [90,91]. The other time period of peak fluxes generally occurs near the time of 50% heading (i.e., approximately the time of anthesis) and has been linked to the sink-source relationship of photosynthates in the plant when CH_4 fluxes have been observed to increase during vegetative growth as root exudates increase and decrease following heading as fixed-C is translocated to developing grain. This plant-related peak has been observed in several studies [15–17,80,92–94,95] and similar seasonal trends have been observed in root growth [96–98], root exudation rates [66], and anaerobic root respiration rates [99].

6.4. Transport mechanisms

The three mechanisms by which CH_4 is transported from a ponded soil to the atmosphere are diffusion through the floodwater, ebullition, and plant-mediated diffusion. Diffusion of CH_4 through overlying floodwater is minor as diffusion of gases is approximately 10,000 times slower through water than through air [46]. Ebullition, bubbles forming and forcing their way to the surface, may be a significant transport mechanism early in the season, especially with high OM inputs, soil disturbances, and in coarse-textured soils, but generally plays only a small role in CH_4 transport, which diminishes as plants mature and plant-mediated transport (PMT) increases [76,100]. The majority of CH_4 emissions from a rice system occur through the rice plants via aerenchyma cells, where studies have indicated that about 90% of season-long emissions are released through the rice plants, compared to 8 to 9% released by ebullition and 1 to 2% by diffusion through the floodwater [100–104].

Based on experiments using artificial atmospheres of various gas compositions, Denier van der Gon and van Breemen [105] determined that PMT is driven by molecular diffusion and not affected by transpiration or stomatal opening. Others have observed a decreasing CH_4 concentration gradient from the soil to the rice root aerenchyma, shoot aerenchyma, and atmosphere, indicative of a diffusive transport pathway from the soil to the atmosphere through the plant [104,106]. Other studies have also confirmed that CH_4 transport is not related to transpiration and is unaffected by cutting plants just above the water surface [103,104,107]. However, Hosono and Nouchi [108] determined that PMT was reduced linearly as roots were cut and increased with root growth up to heading, indicating that the surface area of roots in contact with soil solution is important in determining PMT. Several studies have determined that the most restrictive zone of CH_4 transport through the rice plant is the root–shoot transition zone where dense intercalary meristem cells restrict movement from the root aerenchyma to the shoot aerenchyma [101,105,106,109,110].

It has been postulated that CH_4 in the gaseous form or dissolved in water enters into root aerenchyma, which forms by degeneration of cortical cells between the exodermis and the vascular bundle, where the dissolved CH_4 is gasified and moves by diffusion from the root aerenchyma through the restrictive transition zone into the aerenchyma of the culm and then to the atmosphere [104,107,109]. It has been determined that CH_4 is released from the rice plant mainly through the lower leaf sheaths. Examining the cultivar 'Koshihikari' with a scanning electron microscope, Nouchi et al. [104] and Nouchi and Mariko [107] observed CH_4 release from 4-μm diameter, hook-shaped micropores arranged regularly approximately 80 μm apart on the abaxial epidermis of leaf sheaths as well as from the connections of leaf sheaths to the culm at nodes. Butterbach-Bahl et al. [106] also determined that CH_4 is primarily released through the lower leaf sheaths, however, micropores were not observed in the cultivars 'Roma' or 'Lido'. More research is required to determine differences in CH_4 release from various cultivars. It has been determined that rice cultivars have differences in CH_4 transport capacity, likely in relation to differences in aerenchyma morphology and the root–shoot transition zone [101] and that CH_4 transport capacity increases as soil temperature increases [108]. Research indicates that PMT is the dominant mechanism of CH_4 release from rice soils and that the rate of transport can be influenced by cultivar or environmental conditions.

7. Factors affecting methane emissions from rice

Through numerous research efforts since the 1980s, several factors have been determined to affect CH_4 emissions from rice cultivation. Due to the complex balance of methanogenesis and methanotrophy that determines how much CH_4 escapes the rice system to the atmosphere along with the large variety of cultural and environmental conditions around the globe, there is large variability in the impact of different factors across time and space. There are a few soil, environmental, and plant factors, however, that seem to have somewhat consistent impacts on CH_4 emissions from rice.

7.1. Soil factors affecting methane emissions from rice

Various studies have observed inconsistent results of N fertilizer application on CH_4 emissions including an increase in emissions with added N [85,90,111,112], a decrease in emissions with added N [113,114], or no impact of added N on CH_4 emissions [15,75,115]. Banger et al. [116] conducted a meta-analysis and determined that CH_4 emissions were significantly greater from N-fertilized rice in 98 out of 155 data pairs, indicating that the increase in plant growth and C fixation resulting from N-fertilization generally increases CH_4 emissions. Wang et al. [78] postulated that the effect of urea on CH_4 emissions may be impacted by pH, where it was observed that urea may cause a decrease in emissions in alkaline soils as urea hydrolysis increases soil pH, limiting the neutrophilic methanogens. In acidic soils, however, the increase in pH from urea hydrolysis shifts the soil pH toward neutral and enhances methanogenesis. Research has consistently indicated that ammonium sulfate reduces CH_4 emissions relative to urea application [70,113,116], likely due to the impact of soil acidification and sulfate reduction decreasing the available C substrate for methanogenesis. Similarly, other studies have determined that oxidized Fe [80,117–120] or NO_3^- [120] amendments have the ability to reduce CH_4 emissions. In addition, Lu et al. [121] observed a 19 to 33% reduction in CH_4 emissions with the application of P due to enhanced root growth and root exudation that was measured in the P-deficient treatment.

Multiple studies have indicated no significant correlations between CH_4 emissions and any static soil properties [68,81] or between CH_4 emissions and total soil C [79,80], while readily mineralizable C has been shown to be positively correlated with CH_4 emissions [75,78]. Particle-size distribution is one soil property that has been regularly related to CH_4 emissions as emissions have been positively correlated with soil sand content [78,80,118,119,122] and inversely correlated with soil clay content [71,78,118,119,122,123]. Studies have observed an increase in CH_4 entrapment resulting from increasing clay contents [71,78], and Sass and Fisher [91] attributed the reduction in CH_4 emissions from clay soils to the entrapment and slow movement of CH_4 that allows more CH_4 to be oxidized in aerated zones surrounding roots and at the soil surface. In a laboratory incubation study, Wang et al. [78] observed varying degrees of CH_4 entrapment, even among soils with similar sand and clay contents, where the greatest entrapment (98%) was measured from a Sharkey clay (very-fine, smectitic, thermic Chromic Epiaquerts) soil compared to 81 and 68% entrapment from a Beaumont clay (fine, smectitic, hyperthermic Chromic Dystraquerts) and a Sacramento clay (very-fine, smectitic, thermic Cumulic Vertic Endoaquolls), respectively. This research indicates that clayey soils have the capability of restricting movement of CH_4 to the atmosphere and that other factors, such as clay minerology and soil chemical properties, may impact emissions more than simply the total amount of clay.

7.2. Environmental factors affecting methane emissions from rice

Two major environmental factors that impact CH_4 emissions from rice are temperature and soil saturation status. Numerous studies have observed increases in CH_4 fluxes in relation to increasing soil temperatures [100,108,124]. A study conducted in Japan observed a 1.6-fold increase in emissions from one year to another under the same management and location resulting from an increase in average air temperature from 24.6 to 26.9°C [119]. Methanotrophic activity changes only slightly between 10 and 40°C, while temperature has a strong influence on methanogenesis [56], which leads to a decrease in the proportion of CH_4 oxidized and an increase in emissions as soil temperature increases. Van Winden et al. [65], for example, reported 98% CH_4 oxidation at 5°C compared to 50% oxidation at 25°C.

Soil saturation status has a profound influence on CH_4 emissions through the impact of saturation on soil redox processes, such as methanogenesis. Methane emissions have been observed from soils at an Eh as great as –100 mV [125], while emissions increase as Eh decreases. The amount of time required after saturation to reach low redox potentials condu-cive to methanogenesis varies based on soil textural and chemical properties [119], but generally occurs within several days or weeks after flooding. Studies have indicated that a single mid-season drainage can reduce CH_4 emissions by as much as 65% [68,70,75,95,113,126], however, the potential for greenhouse gas mitigation is reduced or negated due to an increase in N_2O emissions resulting from the drainage [70,113,126,127]. Further research is needed in order to more adequately understand the balance between CH_4 and N_2O emissions under various water management regimes as well as the impact that N management has on emissions when fields are drained.

7.3. Plant factors affecting methane emissions from rice

Due to the strong impact of rice plants on CH_4 transport and CH_4 production from root exudates and residue, there are several plant factors that significantly impact emissions from rice cultivation. A strong relationship between plant growth and CH_4 emissions has been observed in many studies [16,17,80,92–95], particularly in temperate regions, where much of the previous crop's residue decomposes during the winter. Studies have indicated that CH_4 emissions are up to 20 times greater from soil planted with rice than from unvegetated soil [67,107,123], indicating the large influence of rice plants on emissions.

One of the major plant factors impacting CH_4 emissions from rice is whether or not a ratoon crop is grown. This impact is reflected in the USEPA's emissions factors, which are 178 kg CH_4-C ha^{-1} for non-California primary rice crops and an additional 585 kg CH_4-C ha^{-1} when a ratoon crop is produced [2], based on ratoon crops studied in Louisiana [21,22]. The large increase in emissions from ratoon crops is likely a result of large quantities of residue inputs from the harvest of the primary crop in addition to well-developed root systems that further increase the available C for methanogenesis. Lindau et al. [22] observed a significant positive correlation between rice straw additions from a primary crop and resulting emissions from the following ratoon crop.

Another plant factor that has a substantial impact on CH_4 emissions is biomass accumulation. Huang et al. [128] determined that CH_4 fluxes measured during the growing season were positively correlated to aboveground and belowground dry matter on the dates of flux measurements. Additional studies have observed positive correlations between season-long CH_4 emissions and aboveground [16,72,102,128] and belowground dry matter [129]. These studies have indicated a strong relationship between plant growth and CH_4 emissions, which may result from an increase in available substrate as root exudates have been correlated to biomass [66].

Cultivar selection has also been shown to be an important plant factor influencing CH_4 emissions from rice. While the mechanisms for cultivar differences in CH_4 emissions have not been extensively studied, it appears that differences likely arise from variability in CH_4 transport capacity, biomass or dry matter production, root exudation, and microbial community dynamics among cultivars. Butterbach-Bahl et al. [101], for example, attributed a 24 to 31% difference in emissions between two pure-line cultivars to differences in CH_4 transport capacities, as no differences were observed between CH_4 production or oxidation. Aulakh et al. [84] observed a positive correlation between TOC from root exudates and CH_4 production potential, indicating the potential for cultivar differences in emissions based on variable root exudation rates. Previous studies have reported reduced emissions from semi-dwarf relative to standard-stature cultivars [22,91,130]. The difference in CH_4 emissions between semi-dwarf and standard-stature cultivars observed in these studies may be a result of the positive correlation between dry matter and C exudation rates from roots [84] or between aboveground dry matter and CH_4 emissions [16,72,102,128]. While a reduction in emissions from semi-dwarf cultivars is oftentimes linked to reduced dry matter accumulation, Rogers et al. [93] observed a reduction in aboveground dry matter that was not accompanied by a reduction in emissions. Furthermore, Sigren et al. [130] measured greater emissions accompanied by greater soil

Figure 2. Methane emissions from standard-stature, conventional rice varieties, such as "Taggart" (top left) and "Wells" (top right), and hybrids varieties, such as "CLXL745" (bottom) have recently been studied in the field at the Rice Research and Extension Center near Stuttgart, AR. Photographs taken by K. Brye.

acetate concentrations from a standard stature ('Mars') relative to a semi-dwarf cultivar ('Lemont'), while aboveground dry matter was similar between the two cultivars. Huang et al. [128] indicated that, while biomass may explain differences in emissions within one cultivar, the intervarietal differences in biomass are small in comparison to differences in emissions, indicating that another factor besides aboveground dry matter impacts intervarietal differences in CH_4 emissions.

Cultivar differences, however, extend beyond the impact of biomass production on emissions. Ma et al. [131] observed a 67% increase in CH_4 oxidation from a hybrid cultivar accompanied by a reduction in emissions and soil CH_4 concentration relative to pure-line cultivars. Additional studies have also identified 25 to 37% reductions in fluxes from hybrid relative to pure-line cultivars [93,132,133] (Figure 2). This indicates that greater methanotrophic activity in the rhizosphere of hybrid cultivars may reduce CH_4 fluxes by oxidizing a greater proportion of the produced CH_4. It is clear that cultivar selection has the potential for mitigation of CH_4 from rice cultivation. However, due to the lack of understanding the mechanisms for differences in emissions, it appears that direct CH_4 flux measurements from various cultivars are necessary in determining emissions differences until further research clarifies the understanding for cultivar differences in CH_4 emissions (Figure 2).

8. Conclusions

Though some knowledge has been gained, there is much more that still needs to be learned and understood regarding CH_4 emissions from rice production in the US, its contribution to climate change, and potential mitigation strategies. Additional field research needs to be conducted to better assess the magnitudes and relative contributions the various known factors have on CH_4 production and emission from soils used for rice production.

It is possible that a single CH_4 emissions factor for application to all non-California-grown, primary-crop rice in the US is too general. Consequently, the single CH_4 emissions factor may be a severe overestimation for some rice-producing areas, while being an underestimation for other areas. Only after additional data have been generated can regulatory agencies, such as the USEPA, further refine greenhouse gas emissions factors to reflect the large variety of soils and agronomic cultural practices throughout the temperate US and combat the potential negative effects of climate change.

Author details

Alden D. Smartt, Kristofor R. Brye and Richard J. Norman

*Address all correspondence to: kbrye@uark.edu

Department of Crop, Soil, and Environmental Sciences, University of Arkansas, Fayetteville, USA

References

[1] Smith P, Martino D, Cai Z, Gwary D, Janzen H, Kumar P, McCarl B, Ogle S, O'Mara F, Rice C, Scholes B, Sirotenko O. Agriculture. In: Solomon S, Qin D, Manning M, Chen Z, Marquis M, Averyt KB, Tignor M, Miller HL, editors. Climate Change 2007:

The Physical Science Basis. Contribution of Working Group I to the Fourth Assessment Report of the Intergovernmental Panel on Climate Change. Cambridge: Cambridge University Press; 2007. pp. 498–540.

[2] United States Environmental Protection Agency. Inventory of U.S. Greenhouse Gas Emissions and Sinks: 1990–2013 [Internet]. 2015. Available from: http://www3.epa.gov/climatechange/Downloads/ghgemissions/US-GHG-Inventory-2015-Main-Text.pdf (Accessed 2015-09-25.

[3] United States Environmental Protection Agency. Global Anthropogenic Non-CO2 Greenhouse Gas Emissions: 1990–2020 [Internet]. 2006. Available from: http://nepis.epa.gov/ Adobe/PDF/2000ZL5G.PDF (Accessed 2015-04-06).

[4] United States Environmental Protection Agency. Inventory of U.S. Greenhouse Gas Emissions and Sinks: 1990–2012 [Internet]. 2014. Available from: http://www.epa.gov/climatechange/Downloads/ ghgemissions/US-GHG-Inventory-2014-Main-Text.pdf (Accessed 2014-12-12).

[5] Tyndall J. On the absorption and radiation of heat by gases and vapours, and on the physical connection. Philos. Mag. 1861;22:277–302.

[6] Ramanathan V. Greenhouse effect due to chlorofluorocarbons: Climatic implications. Science. 1975;190:50–52.

[7] Forster P, Ramaswamy V, Artaxo P, Berntsen T, Betts R, Fahey DW, Haywood J, Lean J, Lowe DC, Myhre G, Nganga J, Prinn R, Raga G, Schulz M, Van Dorland R. Changes in atmospheric constituents and in radiative forcing. In: Solomon S, Qin D, Manning M, Chen Z, Marquis M, Averyt KB, Tignor M, Miller HL, editors. Climate Change 2007: The Physical Science Basis. Contribution of Working Group I to the Fourth Assessment Report of the Intergovernmental Panel on Climate Change. Cambridge: Cambridge University Press; 2007. pp. 129–234.

[8] Indermühle A, Stocker TF, Joos F, Fischer H, Smith HJ, Wahlen M, Deck B, Mastroianni D, Tschumi J, Blunier T, Meyer R, Stauffer B. Holocene carbon-cycle dynamics based on CO_2 trapped in ice at Taylor Dome, Antarctica. Nature. 1999;398:121–126.

[9] Tans P, Keeling R. Mauna Loa CO_2 Monthly Mean Data [Internet]. 2013. Available from: http://www.esrl.noaa.gov/ gmd/ccgg/trends/#mlo (Accessed 2013-05-15).

[10] Intergovernmental Panel on Climate Change. Climate Change 2007: Mitigation of Climate Change. Contribution of Working Group III to the Fourth Assessment Report of the Intergovernmental Panel on Climate Change. Metz B, Davidson OR, Bosch PR, Dave R, Meyer LA, editors. Cambridge: Cambridge University Press; 2007. 851 p.

[11] United State Department of Agriculture, National Agricultural Statistics Service. Crop Production 2013 Summary [Internet]. 2014. Available from: http://

usda.mannlib.cornell.edu/usda/nass/CropProdSu//2010s/2014/CropProd-Su-01-10-2014.pdf (Accessed 2015-09-25).

[12] Intergovernmental Panel on Climate Change. Volume 4: Agriculture, Forestry and Other Land Use. Chapter 5: Cropland [Internet]. 2006. Available from: http://www.ipcc.ch/meetings/session25/doc4a4b/vol4.pdf (Accessed 2015-09-20).

[13] Byrd GT, Fisher FM, Sass RL. Relationships between methane production and emission to lacunal methane concentrations in rice. Global Biogeochem. Cycl. 2000;14:73–83.

[14] Kongchum M. Effect of plant residue and water management practices on soil redox chemistry, methane emission, and rice productivity. PhD Diss. Louisiana State Univ., Baton Rouge; 2005.

[15] Rogers CW, Brye KR, Norman RJ, Gbur EE, Mattice JD, Parkin TB, Roberts TL. Methane emissions from drill-seeded, delayed-flood rice production on a silt-loam soil in Arkansas. J. Environ. Qual. 2013;42:1059–1069.

[16] Sass RL, Fisher FM, Harcombe PA, Turner FT. Mitigation of methane emissions from rice fields: Possible adverse effects of incorporated rice straw. Global Biogeochem. Cyc. 1991;5:275–287.

[17] Sass RL, Fisher FM, Turner FT, Jund MF. Methane emissions from rice fields as influenced by solar radiation, temperature, and straw incorporation. Global Biogeochem. Cyc. 1991;5:335–350.

[18] Sass RL, Andrews JA, Ding A, Fisher FM. Spatial and temporal variability in methane emissions from rice paddies: Implications for assessing regional methane budgets. Nutr. Cycl. Agroecosys. 2002;64:3–7.

[19] Sass RL, Cicerone RJ. Photosynthate allocations in rice plants: food production or atmospheric methane? Proc. Natl. Acad. Sci. USA 2002;99:11993–11995.

[20] Yao H, Jingyan J, Lianggang Z, Sass RL, Fisher FM. Comparison of field measurements of CH_4 emission from rice cultivation in Nanjing, China and in Texas, USA. Adv. Atmos. Sci. 2001;18:1121–1130.

[21] Lindau CW, Bollich PK. Methane emissions from Louisiana first and ratoon crop rice. Soil Sci. 1993;156:42–48.

[22] Lindau CW, Bollich PK, DeLaune RD. Effect of rice variety on methane emission from Louisiana rice. Agric. Ecosys. Environ. 1995;54:109–114.

[23] Bossio DA, Horwath WR, Mutters RG, van Kessel C. Methane pool and flux dynamics in a rice field following straw incorporation. Soil Biol. Biochem. 1999;31:1313–1322.

[24] Fitzgerald GJ, Scow KM, Hill JE. Fallow season straw and water management effects on methane emissions in California rice. Global Biogeochem. Cycl. 2000;14:767–776.

[25] United States Department of Agriculture, Foreign Agricultural Service. World Agricultural Production [Internet]. 2013. Available from: http://www.fas.usda.gov/psdonline/circulars/ production.pdf (Accessed 2013-07-09).

[26] Chang TT. Rice. In: Kiple KF, Ornelas KC, editors. The Cambridge World History of Food Vol. 1. Cambridge: Cambridge University Press; 2000. pp. 132–148.

[27] Maclean JL, Dawe DC, Hardy B, Hettel GP, editors. Rice Almanac: Source Book for the Most Important Economic Activity on Earth, 3ʳᵈ Ed. Wallingford, United Kingdom: CABI Publishing; 2002. 257 p.

[28] Lu JJ, Chang TT. 1980. Rice in its temporal and spatial perspectives. In: Luh BS editor. Rice: Production and Utilization. Westport, CT: AVI Publishing Co.; 1980. pp. 1–74.

[29] Chang TT. 2003. Origin, domestication, and diversification. In: Smith CW, Dilday RH, editors. Rice: Origin, History, Technology, and Production. Hoboken: Wiley Sciences; 2003. pp. 3–25.

[30] United States Department of Agriculture, Foreign Agricultural Service. Grain: World Markets and Trade [Internet]. 2013. Available from: http://usda01.library.cornell.edu/ usda/fas/grain-market//2010s/2013/grain-market-02-08-2013.pdf   (Accessed 2013-03-08).

[31] Hardke JT, Wilson Jr CE. Trends in Arkansas rice production. In: Norman RJ, Moldenhauer KAK, editors. B.R. Wells Rice Research Studies 2012. Fayetteville; Arkansas Agric. Expt. Stn. Res. Ser. 609; 2013. pp. 38–47.

[32] De Datta SK. 1981. Principles and Practices of Rice Production. New York: John Wiley & Sons; 1981. 642 p.

[33] Chang TT. 1999. The prospect of rice production increase. In: Food Needs of the Developing World in the Early Twenty-first Century. Pontifical Academy of Sciences and Oxford University Press, Oxford.

[34] Street JE, Bollich PK. Rice production. In: Smith CW, Dilday RH, editors. Rice: Origin, History, Technology, and Production. Hoboken: Wiley Sciences; 2003. pp. 271–296.

[35] Martin JH, Leonard WH, Stamp DL, editors. Principles of Field Crop Production. New York: Macmillan; 1976. 1118 p.

[36] Linscombe SD, Saichuk JK, Seilhan KP, Bollich PK, Funderburg ER. General agronomic guidelines. In: Louisiana Rice Production Handbook. LSU Agric. Ctr. Publ. 2321; 1999. pp. 5–12.

[37] Klosterboer AD, Turner FT. Seeding methods. In: 1999 Rice Production Guidelines. Tex. Agric. Ext. Serv. Publ. D-1253; 1999. p. 8.

[38] Moldenhauer K, Wilson Jr CE, Counce P, Hardke J. Rice growth and development. In: Hardke JT, editor. Arkansas Rice Production Handbook. Little Rock: University of

Arkansas Division of Agriculture Cooperative Extension Service MP192; 2013. pp. 9–20.

[39] Hardke JT, Wilson Jr CE. Introduction. In: Hardke JT, editor. Arkansas Rice Production Handbook. Little Rock: University of Arkansas Division of Agriculture Cooperative Extension Service MP192; 2013. pp. 1–8.

[40] Hill JE, Roberts SR, Brandon DM, Scardaci SC, Williams JF, Wick CM, Canevari WM, Weir BL. Rice production in California. Coop. Ext. Univ. Calif. Div. Agric. Nat. Res. Publ. 21498. 1992.

[41] Wilson CE, Wamishe Y, Lorenz G, Hardke J. Rice stand establishment. In: Hardke JT, editor. Arkansas Rice Production Handbook. Little Rock; University of Arkansas Division of Agriculture Cooperative Extension Service MP192; 2013. pp. 31–40.

[42] Wilson CE, Runsick SK, Mazzanti R. Trends in Arkansas rice production. In: Norman RJ, Moldenhauer KAK, editors. B.R. Wells Rice Research Studies, 2009. Fayetteville; University of Arkansas Agric. Expt. Stn. Res. Series 581; 2010. pp. 11–21.

[43] Norman R, Slaton N, Roberts T. Soil fertility. In: Hardke JT, editor. Arkansas Rice Production Handbook. Little Rock: University of Arkansas Division of Agriculture Cooperative Extension Service MP192; 2013. pp. 69–102.

[44] Scott H, Miller D, Renaud F. Rice soils: Physical and chemical characteristics and behavior. In: Smith CW, Dilday RH, editors. Rice: Origin, History, Technology, and Production. Hoboken: Wiley Sciences; 2003. pp. 297–330.

[45] Livingston G, Hutchinson G. Enclosure-based measurement of trace gas exchange: applications and sources of error. In: Matson PA, Harris RC, editors. Biogenic Trace Gases: Measuring Emissions from Soil and Water. Osney Mead, Oxford: Blackwell Sciences Ltd.; 1995. pp. 14–51.

[46] Greenwood DJ. The effect of oxygen concentration on the decomposition of organic materials in soil. Plant Soil. 1961;14:360–376.

[47] Hillel D. Introduction to Environmental Soil Physics. San Diego: Elsevier Academic Press; 2004. 494 p.

[48] Nazaroff WW. Radon transport from soil to air. Rev. Geophy. 1992;30:37–160.

[49] Vepraskas MJ, Faulkner SP. Redox chemistry of hydric soils. In: Richardson JL, Vepraskas MJ, editors. Wetland Soils: Genesis, Hydrology, Landscapes, and Classification. Boca Raton: CRC Press, Taylor & Francis Group; 2001. pp. 85–105.

[50] Turner FT, Patrick Jr WH. Chemical changes in waterlogged soils as a result of oxygen depletion. Trans. 9th Intern. Cong. Soil Sci. 1968;4:53–65.

[51] Brady NC, Weil RR. The Nature and Properties of Soil. 14th ed. Upper Saddle River, New Jersey: Prentice-Hall; 2008. 975 p.

[52] Patrick WH, Gambrell RP, Faulkner SP. Redox measurements of soil. In: Sparks DL, editor. Methods of Soil Analysis, Part 3: Chemical Methods, 3rd Ed. Madison: SSSA; 1996. pp. 1255–1273.

[53] Patrick WH, Jugsujinda A. Sequential reduction and oxidation of inorganic nitrogen, manganese, and iron in flooded soil. Soil Sci. Soc. Am. J. 1992;56:1071–1073.

[54] Conrad R. Control of methane production in terrestrial ecosystems. In: M.O. Andreae and D.S. Schimel, editors, Exchange of trace gases between terrestrial ecosystems and the atmosphere. New York: John Wiley & Sons; 1989. p. 39–58.

[55] Conrad R, Erkel C, Liesack W. Rice cluster I methanogens, an important group of *Archaea* producing greenhouse gas in soil. Curr. Opin. Biotech. 2006;17:262–267.

[56] Nazaries L, Murrell JC, Millard P, Baggs L, Singh BK. Methane, microbes and models: Fundamental understanding of the soil methane cycle for future predictions. Environ. Microbiol. 2013;15:2395–2417.

[57] Grosskopf R, Stubner S, Liesack W. Novel euryarchaeotal lineages detected on rice roots and in the anoxic bulk soil of flooded rice mesocosms. Appl. Environ. Microbiol. 1998;64:4983–4989.

[58] Seedorf H, Dreisbach A, Hedderich R, Shima S, Thauer RK. $F_{420}H_2$ oxidase (FprA) from *Methanobrevibacter arboriphilus*, a coenzyme F_{420}-dependent enzyme involved in O_2 detoxification. Arch. Microbiol. 2004;182:126–137.

[59] Conrad R, Klose M, Noll M, Kemnitz D, Bodelier PLE. Soil type links microbial colonization of rice roots to methane emissions. Glob. Change Biol. 2008;14:657–669.

[60] Ramakrishnan B, Lueders T, Dunfield PF, Conrad R, Friedrich MW. Archaeal community structures in rice soils from different geographical regions before and after initiation of methane production. FEMS Microbiol. Ecol. 2001;37:175–186.

[61] Kruger M, Eller G, Conrad R, Frenzel P. Seasonal variation in pathways of CH_4 production and in CH_4 oxidation in rice fields determined by stable carbon isotopes and specific inhibitors. Global Change Biol. 2002;8:265–280.

[62] Lu YH, Conrad R. In situ stable isotope probing of methanogenic archaea in the rice rhizosphere. Science. 2005;309:1088–1090.

[63] Conrad R. Control of microbial methane production in wetland rice fields. Nutr. Cycl. Agroecosys. 2002;64:59–69.

[64] Bender M, Conrad R. Kinetics of CH_4 oxidation in oxic soils exposed to ambient air or high CH_4 mixing ratios. FEMS Microbiol. Ecol. 1992;101:261–270.

[65] Van Winden JF, Reichart GJ, McNamara NP, Benthien A, Sinninghe Damste JS. Temperature-induced increase in methane release from peat bogs: a mesocosm experiment. PLOS One. 2012;7:e39614.

[66] Aulakh MS, Wassmann R, Bueno C, Kreuzwieser J, Rennenberg H. Characterization of root exudates at different growth stages of ten rice (*Oryza sativa* L.) cultivars. Plant Biol. 2001;3:139–148.

[67] Dannenberg S, Conrad R. Effect of rice plants on methane production and rhizospheric metabolism in paddy soil. Biogeochemistry. 1999;45:53–71.

[68] Lu Y, Wassmann R, Neue HU, Huang C, Bueno CS. Methanogenic responses to exogenous substrates in anaerobic rice soils. Soil Biol. Biochem. 2000;32:1683–1690.

[69] Buendia LV, Neue HU, Wassmann R, Lantin RS, Javellana AM, Arah J, Wang Z, Wanfang L, Makarim AK, Corton TM, Charoensilp N. An efficient sampling strategy for estimating methane emission from rice field. Chemosphere. 1998;36:395–407.

[70] Bronson KF, Neue HU, Singh U, Abao EB. Automated chamber measurements of methane and nitrous oxide flux in a flooded rice soil: Residue, nitrogen, and water management. Soil Sci. Soc. Am. J. 1997;61:981—987.

[71] Denier van der Gon HAC, van Breemen N, Neue HU, Lantin RS, Aduna JB, Alberto MCR, Wassmann R. 1996. Release of entrapped methane from wetland rice fields upon soil drying. Global Biogeochem. Cycl. 1996;10:1–7.

[72] Shang Q, Yang X, Gao C, Wu P, Liu J, Xu Y, Shen Q, Zou J, Guo S. Net annual global warming potential and greenhouse gas intensity in Chinese double rice-cropping systems: a 3-year field measurement in long-term fertilizer experiments. Global Change Biol. 2011;17:2196–2210.

[73] Tsutsuki K, Ponnamperuma FN. Behavior of anaerobic decomposition products in submerged soils: Effects of organic material amendment, soil properties, and temperature. Soil Sci. Plant Nutr. 1987;33:13–33.

[74] Schutz H, Holzapfel-Pschorn A, Conrad R, Rennenberg H, Seiler W. A 3-year continuous record on the influence of daytime, season, and fertilizer treatment on methane emission rates from in Italian rice paddy. J. Geophys. Res. 1989;94:16405–16416.

[75] Yagi K, Minami K. Effect of organic matter application on methane emission from some Japanese paddy fields. Soil Sci. Plant Nutr. 1990;36:599–610.

[76] Denier van der Gon HAC, Neue HU. Influence of organic matter incorporation on the methane emission from a wetland rice field. Global Biogeochem. Cycl. 1995;9:11–22.

[77] Inoko A. 1984. Compost as a source of plant nutrients. In: Organic Matter and Rice. Los Banos, Philippines: International Rice Research Institute; 1984. pp. 137–145.

[78] Wang ZP, Lindau CW, DeLaune RD, Patrick WH. Methane emission and entrapment in flooded rice soils as affected by soil properties. Biol. Fert. Soils. 1993;16:163–168.

[79] Cheng W, Yagi K, Akiyama H, Nishimura S, Sudo S, Fumoto T, Hasegawa T. An empirical model of soil chemical properties that regulate methane production in Japanese rice soils. J. Environ. Qual. 2007;36:1920–1925.

[80] Huang Y, Jiao Y, Zong L, Zheng X, Sass RL, Fisher FM. Quantitative dependence of methane emission on soil properties. Nutr. Cycl. Agroecosys. 2002;64:157–167.

[81] Neue HU, Wassmann R, Kludze HK, Bujun W, Lantin RS. Factors and processes controlling methane emissions from rice fields. Nutr. Cycl. Agroecosys. 1997;49:111–117.

[82] Watanabe A, Takeda T, Kimura K. Evaluation of origins of CH_4 carbon emitted from rice paddies. J. Geophys. Res. 1999;104:23623–23629.

[83] Chidthaisong A, Watanabe I. Methane formation and emission from flooded rice soil incorporated with ^{13}C-labeled rice straw. Soil Biol. Biochem. 1997;29:1173–1181.

[84] Aulakh MS, Wassmann R, Bueno C, Rennenberg H. Impact of root exudates of different cultivars and plant development stages of rice (*Oryza satiza* L.) on methane production in a paddy soil. Plant Soil 2001;230:77–86.

[85] Denier van der Gon HAC, Kropff MJ, van Breemen N, Wassmann R, Lantin RS, Aduna E, Croton TM, van Laar HH. Optimizing grain yields reduces CH_4 emissions from rice paddy fields. P. Natl. Acad. Sci. USA. 2002;99:12021–12024.

[86] Minoda T, Kimura M. Contribution of photosynthized carbon to the methane emitted from paddy fields. Geophys. Res. Lett. 1994;21:2007–2010.

[87] Minoda T, Kimura M, Wada E. Photosynthates as dominant sources of CH_4 and CO_2 in soil water and CH_4 emitted to the atmosphere from paddy fields. J. Geophys. Res. 1996;101:21091–21097.

[88] Ponnamperuma FN. Some aspects of the physical chemistry of paddy soils. In: Institute of Soil Science, Academia Sinica, editor. Proceedings of Symposium on Paddy Soils. Beijing, Science Press; 1981. pp. 59–94.

[89] Wassmann R, Buendia LV, Lantin RS, Bueno CS, Lubigan LA, Umali A, Nocon NN, Javellana AM, Neue HU. Mechanisms of crop management impact on methane emissions from rice fields in Los Banos, Philippines. Nutr. Cycl. Agroecosys. 2000;58:107–119.

[90] Lindau CW, Bollich PK, DeLaune RD, Patrick WH, Law VJ. Effect of urea and environmental factors on CH_4 emissions from a Louisiana, USA rice field. Plant Soil. 1991;136:195–203.

[91] Sass RL, Fisher FM. Methane emissions from rice paddies: a process study summary. Nutr. Cycl. Agroecosys. 1997;49:119–127.

[92] Nouchi I, Hosono T, Aoki K, Minami K. Seasonal variation in methane flux from rice paddies associated with methane concentration in soil water, rice biomass and temperature, and its modeling. Plant Soil. 1994;161:195–208.

[93] Rogers CW, Brye KR, Smartt AD, Norman RJ, Gbur EE, Evans-White MA. Cultivar and previous crop effects on methane emissions from drill-seeded, delayed-flood rice production on a silt-loam soil. Soil Sci. 2014;179:28–36.

[94] Sass RL, Fisher FM, Harcombe PA, Turner FT. Methane production and emissions in a Texas rice field. Global Biogeochem. Cyc. 1990,4:47–68.

[95] Sass RL, Fisher FM, Wang YB, Turner FT, Jund MF. Methane emissions from rice fields: The effect of flood water management. Global Biogeochem. Cyc. 1992;6:249–262.

[96] Beyrouty CA, Wells BR, Norman RJ, Marvel JN, Pillow JA. Root growth dynamics of a rice cultivar grown at two locations. Agron. J. 1988;80:1001–1004.

[97] Beyrouty CA, Norman RJ, Wells BR, Hanson MG, Gbur EE. Shoot and root growth of eight rice cultivars. In: Wells BR, editor. Arkansas Rice Research Studies 1992. Fayetteville; Arkansas Agric. Expt. Stat. Res. Ser. 431; 1993. pp. 119–122.

[98] Slaton NA, Beyrouty CA, Wells BR, Norman RJ, Gbur EE. Root growth and distribution of two short-season rice genotypes. Plant Soil. 1990;121:269–278.

[99] Tolley MD, DeLaune RD, Patrick WH. The effect of sediment redox potential and soil acidity on nitrogen uptake, anaerobic root respiration, and growth of rice (Oryza sativa). Plant Soil. 1986;93:323–331.

[100] Schutz H, Seiler W, Conrad R. Processes involved in formation and emission of methane in rice paddies. Biogeochemistry. 1989;7:33–53.

[101] Butterbach-Bahl K, Papen H, Rennenberg H. Impact of gas transport through rice cultivars on methane emission from rice paddy fields. Plant Cell Environ. 1997;20:1175–1183.

[102] Cicerone RJ, Shetter JD. Sources of atmospheric methane: Measurements in rice paddies and a discussion. J. Geophys. Res. 1981;86:7203–7209.

[103] Holzapfel-Pschorn A, Conrad R, Seiler W. Effects of vegetation on the emission of methane from submerged paddy soil. Plant Soil. 1986;92:223–233.

[104] Nouchi I, Mariko S, Aoki K. Mechanism of methane transport from the rhizosphere to the atmosphere through rice plants. Plant Physiol. 1990;94:59–66.

[105] Denier van der Gon HAC, van Breemen N. Diffusion-controlled transport of methane from soil to atmosphere as mediated by rice plants. Biogeochemistry. 1993;21:177–190.

[106] Butterbach-Bahl K, Papen H, Rennenberg H. Scanning electron microscopy analysis of the aerenchyma in two rice cultivars. Phyton. 2000;40:43–55.

[107] Nouchi I, Mariko S. Mechanisms of methane transport by rice plants. In: Oremland RS, editor. Biogeochemistry of Global Change: Radiatively Active Trace Gases. New York: Chapman & Hall; 1993. pp. 336–352.

[108] Hosono T, Nouchi I. The dependence of methane transport in rice plants on the root zone temperature. Plant Soil, 1997;191:233–240.

[109] Aulakh MS, Wassmann R, Rennenberg H, Fink S. Pattern and amount of aerenchyma relate to variable methane transport capacity of different rice cultivars. Plant Biol. 2000;2:182–194.

[110] Groot TT, van Bodegom PM, Meijer HAJ, Harren FJM. Gas transport through the root-shoot transition zone of rice tillers. Plant Soil. 2005;277:107–116.

[111] Aerts R, Ludwig F. Water table changes and nutritional status affecting trace gas emissions from laboratory columns of peatland soils. Soil Biol. Biochem. 1997;29:1691–1698.

[112] Aerts R, Toet S. Nutritional controls of carbon dioxide and methane emissions from *Carex*-dominated peat soils. Soil Biol. Biochem. 1997;29:1683–1690.

[113] Cai Z, Xing G, Yan X, Xu H, Tsuruta H, Yagi K, Minami K. Methane and nitrous oxide emissions from rice paddy fields as affected by nitrogen fertilisers and water management. Plant Soil. 1997;196:7–14.

[114] Wang Z, DeLaune RD, Lindau CW, Patrick WH. Methane production from anaerobic soil amended with rice straw and nitrogen fertilisers. Fert. Res. 1992;33:115–121.

[115] Adviento-Borbe MA, Pittelkow CM, Anders M, van Kessel C, Hill JE, McClung AM, Six J, Linquist BA. Optimal fertilizer nitrogen rates and yield-scaled global warming potential in drill seeded rice. J. Environ. Qual. 2014;42:1623–1634.

[116] Banger K, Tian H, Lu C. Do nitrogen fertilizers stimulate or inhibit methane emissions from rice fields? Global Change Biol. 2012;18:3259–3267.

[117] Furukawa Y, Inubushi K. Effect of application of iron materials on methane and nitrous oxide emissions from two types of paddy soils. Soil Sci. Plant Nutr. 2004;50:917–924.

[118] Mitra S, Wassmann R, Jain MC, Pathak H. Properties of rice soils affecting methane production potentials: 1. Temporal patterns and diagnostic procedures. Nutr. Cycl. Agroecosys. 2002;64:169–182.

[119] Watanabe A, Kimura M. Influence of chemical properties of soils on methane emission from rice paddies. Comm. Soil Sci. Plant Anal. 1999;30:2449–2463.

[120] Yu KW, Wang ZP, Chen GX. Nitrous oxide and methane transport through rice plants. Biol. Fertil. Soils. 1997;24:341–343.

[121] Lu Y, Wassmann R, Neue HU, Huang C. Impact of phosphorus supply on root exudation, aerenchyma formation and methane emission of rice plants. Biogeochemistry. 1999;47:203–218.

[122] Sass RL, Fisher FM, Lewis ST, Turner FT, Jund MF. Methane emission from rice fields: Effects of soil properties. Global Biogeochem. Cyc. 1994;8:135–140.

[123] Brye KR, Rogers CW, Smartt AD, Norman RJ. Soil texture effects on methane emissions from direct-seeded, delayed-flood rice production in Arkansas. Soil Sci. 2013;178:519–529.

[124] Wang B, Neue HU, Samonte HP. The effect of controlled soil temperature on diel CH_4 emission variation. Chemosphere. 1997;35:2083–2092.

[125] Hou AX, Chen GX, Wang ZP, Van Cleemput O, Patrick Jr WH. Methane and nitrous oxide emissions from a rice field in relation to soil redox and microbiological processes. Soil Sci. Soc. Am. J. 2000;64:2180–2186.

[126] Zou J, Huang Y, Jiang J, Zheng X, Sass RL. A 3-year field measurement of methane and nitrous oxide emissions from rice paddies in China: Effects of water regime, crop residue, and fertilizer application. Global Biogeochem. Cyc. 2005;19:GB2021.

[127] Kreye C, Dittert K, Zheng X, Zhang X, Lin S, Tao H, Sattelmacher B. Fluxes of methane and nitrous oxide in water-saving rice production in north China. Nutr. Cycl. Agroecosyst. 2007;77:293–304.

[128] Huang Y, Sass RL, Fisher FM. Methane emission from Texas rice paddy soils. 2. Seasonal contribution of rice biomass production to CH_4 emission. Global Change Biol. 1997;3:491–500.

[129] Whiting GJ, Chanton JP. Primary production control of methane emission from wetlands. Nature. 1993;364:794–795.

[130] Sigren LK, Byrd GT, Fisher FM, Sass RL. Comparison of soil acetate concentrations and methane production, transport, and emission in two rice cultivars. Global Biogeochem. Cyc. 1997;11:1–14.

[131] Ma K, Qiu Q, Lu Y. Microbial mechanism for rice variety control on methane emission from rice field soil. Global Change Biol. 2010;16:3085–3095.

[132] Simmonds MB, Anders M, Adviento-Borbe MA, van Kessel C, McClung A, Linquist BA. Seasonal methane and nitrous oxide emissions of several rice cultivars in direct-seeded systems. J. Environ. Qual. 2015;44:103–114.

[133] Smartt AD, Rogers CW, Brye KR, Norman RJ, Smartt WJ, Hardke JT, Frizzell DL, Casteneda-Gonzalez E. Growing-season methane fluxes and emissions from a silt-loam soil as influenced by rice cultivar. In: Norman RJ, Moldenhauer KAK, editors. B.R. Wells Rice Research Studies, 2014. Fayetteville; Arkansas Agric. Expt. Stn. Res. Ser. 62; 2015. pp. 289–297.

3

Carbon Dioxide Geological Storage (CGS) – Current Status and Opportunities

Kakouei Aliakbar, Vatani Ali, Rasaei Mohammadreza and Azin Reza

Additional information is available at the end of the chapter

Abstract

Carbon dioxide sequestration has gained a great deal of global interest because of the needs and applications of mitigation strategy in many areas of human endeavors including capture and reduction of CO_2 emission into atmosphere, oil and gas enhanced production, and CO_2 geological storage. In recent years, many developed countries as well as some developing ones have extensively investigated all aspects of the carbon dioxide geological storage (CGS) process such as the potential of storage sites, understanding the behavior of CO_2, and its interaction with various formations comprising trapping mechanisms, flow pattern, and interactions with formation rocks and so on. This review presents a summary of recent research efforts on storage capacity estimation techniques in most prominent storage options (depleted oil and gas reservoir, saline aquifers and coal beds), modeling and simulation means followed by monitoring and verification approaches. An evaluation of the more interesting techniques which are gaining attention in each part is discussed.

Keywords: carbon dioxide, geological storage, CGS

1. Introduction

Carbon dioxide (CO_2) is one of the most emitted greenhouse gases (GHG) which causes heat trapping of the earth and contributes to the global climate change. This global issue led to the public concern and has become a serious problem in the developed and developing countries [1]. Accordingly, the increase of GHG in the atmosphere has led to a rise in the average global temperatures with a warming forecast of 1.8–4.0°C [2]. Recent surveys conducted, see [2–5], show that the CO_2 concentrations has risen from pre-industrial levels of 280 parts per million (ppm) to present levels of ~380 ppm in the atmosphere and this increase in CO_2 concentration depends on world's expanding use of fossil fuels. Further studies, according to the CO_2

emissions from fossil fuel power plants, represent the amount of emissions around 23 Gton-CO_2 per year and 26% of the total emissions approximately[1, 2, 6]. Reports from on-road transportation emissions also indicate the high contribution of CO_2 in atmosphere especially in urban areas. It contributes around 10% of the total global and 20% of the European atmospheric CO_2 emissions [7]. Based on the Intergovernmental panel on Climate Change (IPCC) report in 2005, 72% of the anthropogenic greenhouse effect is due to the CO_2 emission and it is considered as the most important GHG contributor [1]. The Kyoto Protocol in 1997 also recommends that the nations minimize their CO_2 emissions up to 95% of 1990 levels by 2012. In this regard, the mitigation options of the CO_2 have been defined in many national and international scales and the scientists have been looking and developing for the techniques which reduce the CO_2 emissions [8–11]. The options include reduction in using carbon-intensive fuels and improving energy efficiency in order to decrease the CO_2 emissions into the atmosphere or carbon sequestration.

CO_2 sequestration is the process of injecting CO_2 into sub-surface to reduce the emissions of anthropogenic CO_2. According to the IPCC 2005, the storage options are classified into three groups: (1) ocean storage, (2) mineralization, and (3) geological storage. Ocean storage consists of injecting the CO_2 into deep oceans and immobilizing it by dissolving or forming a plume which is heavier than water under the ocean. The ocean is the largest storage option of CO_2 and can contain 40000 Gton of carbon in contrast to the 750 Gton in the atmosphere. The ocean storage has not yet been considered as a pilot scale since it is still in the research phase and may also have dire consequences in marine life in case of leakage during and after the storage. [1, 12]. Mineralization process provides an opportunity to store the CO_2 for a long period of time without any special concern about the permanent mitigation quality. It includes the CO_2 conversion to a solid inorganic carbonates which is stable for a long time. The only considerable problem in this process is related to the high cost of implementation [13]. The CO_2 geological storage (CGS) is considered as the main process for CO_2 sequestration in the developed world [14–16]. The candidate CO_2 storage facilities consist of deep saline aquifer and unmineable coal deposits, as well as depleted and mature oil and gas reservoirs which can contain 2200 Gton of carbon dioxide [17]. Based on an estimation reported by the European technology platform for zero emission fossil fuel power (ZEP), the contribution of each option for the storage potential of CO_2 is shown in Figure1. [18]

As for CGS's regulation in Europe in 2009, the European Union approved that seven million tons of CO_2 could be stored by 2020 and up to 160 million tons by 2030, assuming a 20% reduction in GHG emissions by 2020 [19]. Over the past decade, many developed countries have extensively investigated the potential of CO_2 storage sites as well as understanding the behavior of CO_2 and its interaction with different reservoir formations as a prerequisite to increase the effectiveness and integrity of the CGS projects. These comprise advanced scientific knowledge about CO_2 behavior such as trapping mechanisms (physical and chemical), flow patterns, and interactions with formation rocks that can be achieved by improved techniques such as flow simulation, reservoir modeling, reservoir monitoring, and verification [20].

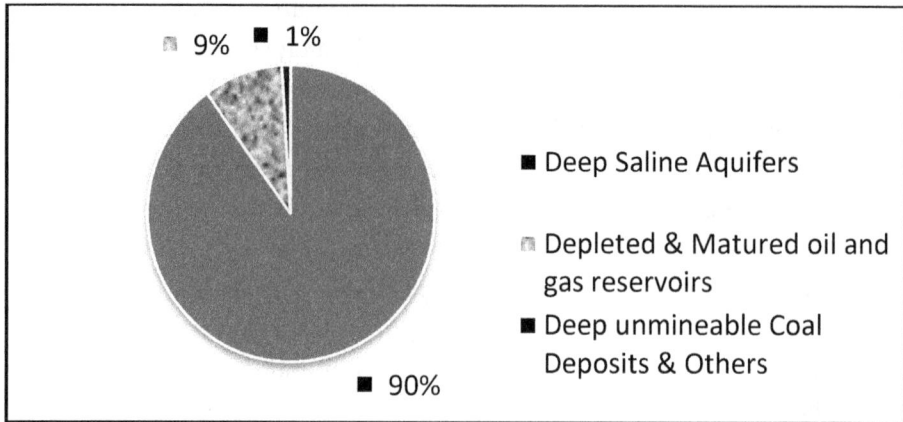

Figure 1. The contribution of most important CGS techniques in the world's CGS projects.

2. CGS: Storage Capacity

In recent years, there have been a number of surveys related to the storage capacity estimation methods in CGS fields [21]. The first groups of estimation assessments were simple with no technical component similar to the estimations held in Europe by Holloway and van der Straaten, in 1995, while the other recent ones have taken into account the complexities and more sophisticated methods of estimating the CO_2 storage capacity [22–28]. One should keep in mind that the capacity estimation in any different scale (global, reservoir, basin, or region) and time frame is a difficult process due to our lack of knowledge about subsurface in most areas of the world and also the uncertainties and inaccessibility of the available data [29]. However, there is a wide variety of estimation techniques proposed by different authors (CSLF, IPCC, and Bradshaw et al.) which mainly rely on a simple algorithm depending on various storage mechanism [26, 28, 30].

In 1979 and 1988, the concept of resource pyramids was developed by Masters and McCabe for the first time and was later proposed to demonstrate the accumulation and quality of the CO_2 storage potentials in the form of three pyramids as an important factor for capacity estimation, including (1) high level, (2) techno-economic, and (3) trap-type and effectiveness pyramid [31, 32]. This concept consists of the main aspects of CO_2 storage such as different time scales and assessment scales, various assessment types, and different geological storage options [29]. For instance, as it has been demonstrated in Figure2, the techno-economic resource pyramid calculates the storage capacity in mass instead of the volume and includes the maximum upper limit of capacity estimate with various time and assessment scales. On the other hands, it reveals three levels of theoretical, realistic and viable estimates in which the theoretical portion includes the entire pyramid whereas the realistic and viable parts have covered the top two portions and only the top portion of pyramid respectively [28, 30].

In an investigation which was performed by Kopp et al. in 2009, to estimate the effective storage capacity, some models were proposed by authors, including(1) CSLF model (proposed by Bachu et al. in 2007 in which the effective storage volume is calculated by reducing the capacity

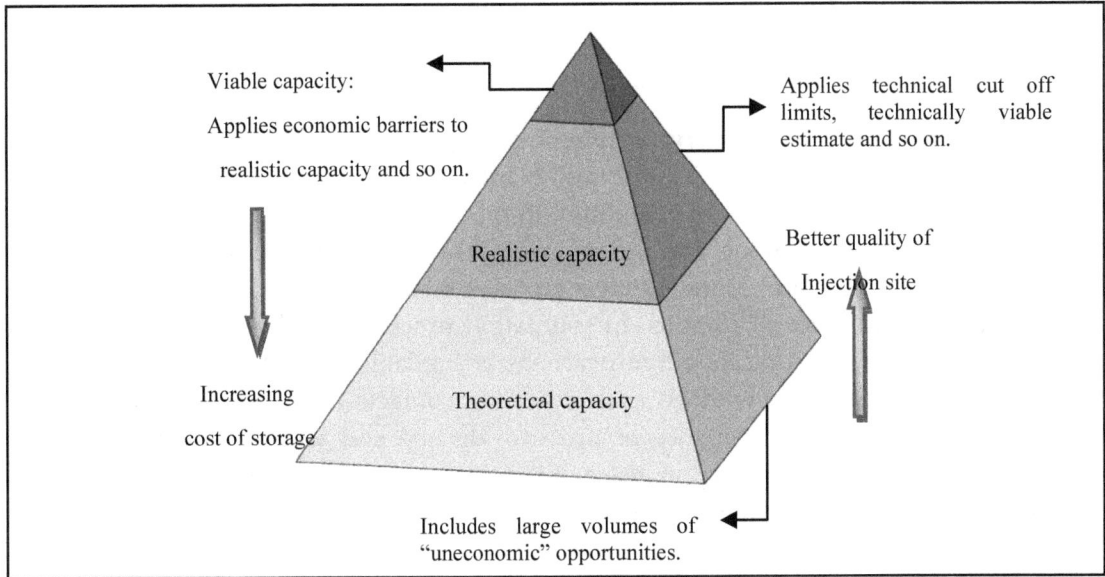

Figure 2. Techno-Economic Resource Pyramid for capacity estimation in CO_2 geological storage.

coefficient from theoretical capacity), (2) Doughty model (proposed by Doughty et al. in 2001 which estimated the effective capacity as a volume fraction for CO_2 storage), and (3) Kopp model (based on Doughty model while the pores containing dissolved CO_2 is much larger than those containing free gas [33]).

According to CO_2 storage capacity estimation surveyed by Bachu et al., based on a summary of carbon sequestration leadership forum (CSLF), different timeframes and field scales are accounted considering various trapping mechanisms (physical and chemical mechanisms) [26]. Bachu et al. have demonstrated the approaches based on different geological potential with generally assessing the opportunity of other storage options like man-made underground cavity and the basalts such as Deccan Plateau in India; however, they need more investigations.

2.1. Estimation techniques in depleted oil and gas reservoir

DOE (2006), 'Methodology for development of carbon sequestration capacity estimates' and CSLF (2007), 'Estimation of CO_2 storage capacity in geological media – phase II' are the major investigations regarding the storage capacity estimation approaches in geological formations. The CSLF (2007) employs a techno-economic resource pyramid in the capacity estimation process for depleted oil and gas reservoir based on McCabe (1998), while the DOE (2006) utilizes volumetric equations and Monte Carlo approach to estimate the uncertainty and capacity storage by incorporating various trapping mechanisms in depleted oil and gas reservoirs [31]. Another integration of DOE and CSLF with simple version of SPE (Society of Petroleum Engineering) petroleum resource management system is proposed and called CO2CRC storage capacity classification [34, 35]. They have reported that on account of greater amount of data in term of oil and gas fields, the estimation process is the easiest among the geological formations. It should be noted that the other methods which are employed in saline aquifers can be used here for CO_2 storage volume estimation: 'volumetric-based estimation' and 'production-based estimation' [35, 36].

Bachu et al., provided a good overview of storage capacity estimates in oil and gas reservoirs to compare the other geological formation such as coal beds and saline aquifers [26]. Based on Bachu et al., the capacity estimation in oil and gas reservoirs is more convenient than other geological formations, and these geological formations are discrete in contrast to the continuous coal beds and saline aquifers [26]. Estimation of the CO_2 storage capacity is also difficult for a number of reasons: In estimation process, some assumption would be made, such as volume occupied by hydrocarbons is available for CO_2 after production for pressure-depleted reservoirs with no hydrodynamic contacts. On the other hand, formation water influx as the consequence of pressure decline and water trapping can be inversed due to the CO_2 injection and increase in the pore spaces which may cause some pores to be unavailable for CO_2 storage. Thus, the original reservoir pressure has the maximum limitation for CO_2 injection into the depleted reservoirs [37]. According to the volume of original oil and gas at surface conditions, theoretical mass storage capacity can be accounted through an equation proposed by Bachu et al.[26]. They also provided an extrapolation to account the theoretical storage capacity in another correlation. In some cases, the actual volume availability to CO_2 storage can be reduced and would be stated by capacity coefficient (equation expressed by Doughty and Press, 2004) [38]. But based on Bachu and Shaw, in 2005, enough data are not available for assessing these coefficients, and estimations are mostly carried out by numerical simulations [9, 38]. One of the specific issues in CO_2 storage in depleted reservoirs is CO_2 flood-enhanced oil recovery. Because of some reasons, the capacity estimation in this case is already an effective estimation. The promising storage sites for CO_2 enhanced recovery can be performed at regional and basin scales such that this criterion decreases the effective capacity to practical storage capacity [39–41].

2.2. Estimation techniques in saline aquifers

As it has been illustrated in recent studies, deep saline aquifers are the most favorable storage option in comparison to the depleted reservoirs and coal beds [1, 27, 28, 39]. In contrast, the numbers of projects which have been conducted by the industries are not considerable due to some reasons, including availability of anthropogenic CO_2 and the related data, site assessment difficulties, poor injectivities, and high cost of monitoring [42]. According to the DOE, a volumetric equation is proposed to CO_2 storage estimation in saline aquifers, while each type of trapping mechanisms is also needed for calculation of the basin-scale assessments [35]. In CSLF methodology for deep saline aquifers, storage estimations based on structural and stratigraphic trapping mechanisms are similar to depleted oil and gas reservoirs, whereas the mass of CO_2 related to the effective storage volume would be more difficult to calculate. Moreover, the storage estimation based on solubility trapping at the basin and regional scales can be calculated by the relation proposed by Bachu and Adams [36, 41].

Bachu et al. proposed a theoretical approach to CO_2 storage estimation considering each type of trapping mechanism in deep saline aquifers [26]. They introduced a simple time-independent volumetric equation used for depleted oil and gas reservoirs in which the traps have been saturated by water rather than being occupied with hydrocarbons. Similar to equation mentioned above, a relation related to the CO_2 mass storage limitation also has developed here for basin- and regional-scale assessments, which can be utilized for theoretical and effective capacity estimations. For residual gas trapping method, the storage volume can be calculated with a time-dependent equation proposed by the authors with regard to the concept of actual CO_2 saturation at flow reversal by Juanes et al. [43]. The solubility mechanism is a time-dependent, continuous, and slow process which can be performed effectively after finishing the injection process. If this trapping system occurs in thick and high permeable aquifers, a convection cell can be constituted and the dissolution process will be improved, while in the case of thin aquifers, this mechanism is less efficient [44, 45]. Capacity storage at the basin and regional scale can be assessed through an equation proposed by Bachu and Adams whereas at the local and site scale, numerical simulation is required for precise estimation of the storage capacity [41]. Estimation through mineral trapping cannot be applied at the regional and basin scales due to the lack of available data and the complex intrinsic of mineral trapping and the chemical and physical related mechanisms. The only remaining approach is numerical simulation which is suitable for site and local scale during a long period of time. According to recent research, mineral trapping mechanism can be compared to the solubility mechanisms with regard to the long time period required here [46, 47]. Hydrodynamic trapping mechanism consists of all the mentioned features of the mechanism and it needs various time scales for acting. This process cannot be evaluated at regional and basin scale estimations due to the different acting time scales through various trapping mechanisms. Hence, it should be considered in a specific point of time and the numerical simulation applied to estimate the storage capacity at local and site scales [26, 48].

De Silva and Ranjith conducted a complete investigation related to the CO_2 estimation methods on saline aquifers and assessed different aspects of the estimation process such as operating time frame, resource circles (pyramids), and trapping mechanisms and factors affecting the storage capacity [50]. The proposed equations in each trapping system are based on the relations recommended by Bachuet al. [26]. The evaluated parameters which can affect the storage capacity consist of *in-situ* pressure, injectivity, temperature, permeability, and compressibility. According to De Silva and Ranjith, eight methods have been introduced to estimate theoretical and effective capacity of CO_2 storages (volumetric method, compressibility method, flow simulation, flow mathematical models, dimensional analysis, analytical investigation, Japanese methodology, and Chinese methodology), while to calculate the practical and matched capacities, the local conditions need to be considered [26, 49, 50]. In a quick and simple volumetric method, the porosity, area, thickness, and storage efficiency of the storage reservoirs are important in capacity estimation according to an equation mentioned by DOE and Ehlig-Economides and Economides [see 51, 52], while van der Meer and Yavuz have proposed another equation to measure the CO_2 mass [53]. To calculate the volume of CO_2 per volume of the aquifers, Eccles et al. have introduced another relation including measuring the effective capacity storage at a special depth [54]. The more comprehensive equation to calculate the storage capacity by compressibility method was shown by

Zhou et al. [55]. The most effective method to assess the capacity is the flow simulation which includes volumetric formulas and more reservoir parameters rather than other methods [56]. Mass balance and constitutive relations are accounted in mathematical models to capacity assessment and dimensional analysis consists of fractional flow formulation with dimensionless assessment and analytical approaches [33]. From the formulations demonstrated by Okwen and Stewart for analytical investigation, it can be deduced that the CO_2 buoyancy and injection rate have affected the storage capacity [57]. Zheng et al. have indicated the equations employed in Japanese and Chinese methodology and have noted that some parameters in Japanese relation can be compared to the CSLF and DOE techniques [58].

2.3. Estimation techniques in coal beds

According to the IPCC 2005, the coal bed storage process is currently in the demonstration phase. MacDonald of Alberta Energy reported the storage in coal bed in 1991 for the first time [59]. One of the most prominent factors to guarantee the successful economic CO_2 storage process is the permeability of coal and it should be more than 1 mD (miliDarcy) [60]. The main problem in CO_2 storage in coal bed process is the limitation of available data about location and capacity of promising sites [30, 26, 28]. It should be noted that the main trapping mechanism in storage process regarding the coal beds is adsorption, and it is necessary to assess the rank, grade, and type of the coal in order to achieve more information about adsorption capacity of the coals [35].

The CSLF and DOE proposed models such as volumetric equation to estimate the coal capacity through substituting the intrinsic methane by injected CO_2 process. Bachu et al. have reported the relation demonstrating the initial gas in place after coal adsorption process proposed by van Bergen et al. and White et al. [59, 61, 62]. One should keep in mind is that since the adsorption is one of the main parts of the storage process, adsorbed gas capacity estimation is also important to investigate [63]. Langmuir equation is a simple and efficient relation for single-layer adsorption capacity estimation in low-pressure conditions [64–66]. In case of high pressure and high temperature, other methods are more suitable such as Bi Langmuir, extended Langmuir, Sips, Langmuir-Freundlich, Toth, UNILAN, two-dimensional EOS, LRC (loading ratio correlation), Dubinin-Radushkevich (D-R) and Dubinin-Astakhov (D-A) [59, 67–73]. A modified Langmuir and Toth correlation was expressed by Himeno et al. and Bae and Bhatia, which includes the substitution of pressure by fugacity high dense phase conditions [74, 75]. Another mathematical power equation proposed by Saghafi et al. can be used to estimate the adsorption capacity [66].

Storage capacity estimation for the stored gas content can be performed through the equation suggested by White, van Bergen et al., CSLF, and Vangkilde et al. [61, 76, 77]. Palarski and Lutynski expressed another relation to estimate the CO_2 storage components in coal seams [78]. To estimate the large-scale storage capacity of 45 important coal basins during Enhanced Coal Bed Methane Recovery (ECBM) in China, Li et al. used an equation which can be modified to a simpler form without considering the different coal bed basins [63, 79].

3. CGS: Modeling and Simulation

To study the behavior of CO_2 during and after the CGS process, numerical modeling is considered as the only effective tool prior to the experimental and field demonstrations instead of analytical and semi-analytical solutions on account of some limitations and simplifications [80–83]. In the past few years, various numerical modeling and reservoir simulations approaches have been documented in the literature at the pilot and commercial scales which are using common numerical methods such as finite difference, finite element, and finite volume methods. One of the most efficient means for reservoir modeling is TOUGH2 simulator developed by Pruess et al. and used successfully in Rio Vista reservoir. In this study, an extension of EOS7R and EWASG modules have been developed to simulate the gas and water flow called EOS7C [84-88]. Omambia and Li carried out a CO_2 numerical modeling in a deep saline aquifer (Wangchang basin, China) using a fluid/property module of TOUGH2 called ECO2N which is adapted from EWASG module [89]. This module was evaluated in a separate study for the CGS process in saline aquifers by Pruess and Spycher [86, 90]. TOUGHREACT, a non-isothermal reactive geochemical transport code, was utilized to simulate the CO_2 disposal in deep aquifers by Xu et al., which was performed by merging the reactive chemistry term into the TOUGH2 framework [91–95]. An efficiency evaluation of CGS was performed in Frio brine pilot project using the TOUGH2 simulator to identify the uncertainties related to nature of the earth by Hovorka et al. [96]. In a previous study at the University of Stuttgart, the MUFTE-UG simulator has been evaluated for CO_2 sequestration in various fields of application such as simulation, CO2SINK, and CO2TRAP [97, 98]. At the Ketzin CO_2 storage site, the ECLIPSE 100/300 and MUFTE-UG codes were employed to perform a history matching [99]. Pawar et al. have investigated a preliminary study to model and simulate the CGS in a depleted oil reservoir by ECLIPSE 100 [100]. Another 2/3 dimensional simulation survey with consideration of reactive flow and transport in deep saline aquifers has been performed by Kumar et al. with GEM simulator (computer modeling groups) [101]. ECLIPSE and DuMux simulators are also taken into consideration to understand the thermal effect during CO_2 injection and movement in the porous medium.

According to the CGS simulation methods, there have been some comparative investigations between the various simulators, such as reported by David et al. and Jiang [102]. David et al. have compared six simulators for numerical simulation of CGS in coal beds: (1) GEM, (2) ECLIPSE, (3) COMET2, (4) SIMED II, (5) GCOMP, and (6) METSIM 2. Additional features are needed to be taken into consideration based on Law et al., such as coal matrix swelling, diffusion of mixed gas, non-isothermal effect, water movement, and so on [103]. According to the recent survey by David et al. GEM and SIMED II are suitable to consider multi-component liquids while ECLIPSE and COMET 2 can handle only two component fluids [103, 104]. In 2011, Jiang demonstrated an overview of the various simulator applications and their numerical features including TOUGHREACT, MUFTE, GEM, ECLIPSE, DuMux, COORES, FEHM, ROCKFLOW, SUTRA, and other types of simulators. Numerical methods and physical models play an important role in the simulators outcomes. Selecting the best simulator among those presented above is highly based on the desired application. For example, the ELSA simulator can be applied efficiently in semi-analytical estimation of fluid distributions; ROCKFLOW is suitable in the case of multi-phase flow and solute transport modeling; GEM is an aqueous geochemistry tool while for the low-temperature situation PHREEQC is more applicable; and for the multi-component, three phase, and 3D fluid flow simulation with consideration of reservoir heterogeneities, COORES would be a robust means [85, 102, 104, 105]. Zhang et al. had a quick look on different types of simulators mentioned earlier and have suggested a new parallel multi-phase fluid flow simulator for CGS in saline aquifers called TOUGH+CO_2 which has been developed on the basis of a modified TOUGH2 family of cods, TOUGH+ and TOUGH2-MP including all the ECO2N features capabilities [83]. This brand new simulator has proved to be a successful and robust means, which has been used in a number of large-scale simulation projects [106–113].

Another group of surveys has focused on the direct modeling of some effective transport phenomena which are essential for predicting parameters that have an important role in underground gas sequestration process such as diffusivity and convection. Azin et al., in 2013, have conducted study regarding correct measurement of diffusivity coefficient [114]. The modeling was based on a method proposed by Sheika et al. to analyze pressure decline data and the impact of pressure and temperature on the measurement of diffusivity coefficient [114]. GholamiY., et al., in 2015, have also investigated the measurement of CO_2 diffusivity in synthetic and saline aquifer solutions at reservoir conditions with emphasis on the role of ion interactions [114–117]. A non-iterative thermodynamic predictive model has investigated by Azin et al. to calculate the effect of gas solubility [118–120]. The effects of convective dissolution and diffusivity mixing have also been surveyed with finite-element method by GholamiY., et al. They have used Streamline Upwind Petrov-Galerkin (SUPG) method and crosswind artificial diffusion and found that the dissolution is controlled by convective dissolution in bulk water [115, 121]. Another numerical simulation was done by Azin et al. to predict the onset of instability in CO_2 underground injection [114]. It was found that depending on Rayleigh number, there is a wave number at which instability occurs earlier and grows faster [114].

4. CGS: Monitoring and Verification

Precise monitoring and verification is required to have an appropriate risk management strategy for the CGS projects [1]. The monitoring and verification process should be commenced from site selection and characterization followed by atmospheric and remote sensing, near and deep surface methods, as well as well bore-monitoring techniques. Different types of monitoring tools are introduced and used in recent literature: acoustic velocity structure imaging by seismic, density distribution imaging by gravity, electrical resistivity structure imaging, and fluid content imaging of potential reservoir rocks by the electromagnetic methods [20, 122]. After injecting the CO_2 into the sequestration sites, electromagnetic and gravitation sensors are employed for seismic surveys of storage integrity such as CO_2 flow and transportation quality in porous media and behavior of cap rock in contact to the CO_2. The leakage measurement in atmospheric level can be done by open path, flux tower, and InSAR systems (satellite-based infrared and interferometric synthetic aperture radar) [20].

Otway Basin Pilot project in Australia is the first CGS project in which monitoring techniques were used [122]. In 2010, the CSEM have considered landing base imaging and passive magnetotelluric in deep crustal scales surveys by Sreitch and colleagues [124]. According to the surveys performed by Arts et al. and Chadwick et al., the 4D gravity and seismic techniques have been successfully accomplished in Sleipner site [125–127]. The 4D vertical seismic profiling (VSP) has been commonly used to quantitative monitoring of the CO_2 plume with tracer injection, well logging, micro-seismic and pressure–temperature measurements which is applied successfully at Frio and Nagaoka project [128–144]. In Frio Brine and Otway Pilot projects, tracer monitoring has been employed to assess the CO_2 breakthrough [145, 146]. The Eddy covariance and hyperspectral imaging in a shallow subsurface site are important computational issues that were examined to monitor the CO_2 leakage in Montana [147, 148]. Another successful surface monitoring technique tested at In Salah project was InSAR which was incorporated into other monitoring techniques such as seismic, gravity, and electromagnetic [149–153]. At Ketzin sequestration site, the monitoring methods included cross-hole resistivity, seismic, and microbiology with temperature and pressure profiling [154-160].

5. Conclusions

In summary, the methods of theoretical and effective capacity estimation of CO_2 storage comprise volumetric and compressibility methods, flow mathematical and simulation models, dimensional analysis, analytical investigation and Japanese/Chinese methodology.

The CSLF model employs a techno-economic resource pyramid in the capacity estimation process for depleted oil and gas reservoir, while the DOE model utilizes volumetric equations and Monte Carlo approach by incorporating various trapping mechanisms. According to the CO2CRC, storage capacity classification in terms of oil and gas fields is the easiest among the other geological options due to the greater amount of data. A volumetric equation has been proposed to CO_2 storage estimation in the most favorable storage option (saline aquifers) while each type of trapping mechanism is also needed for calculation of the basin-scale assessments. The CSLF methodology has been considered for deep saline aquifers as well as depleted oil and gas reservoir based on structural and stratigraphic trapping mechanisms. Estimation through mineral trapping cannot be applied at the regional and basin scales due to lack of data availability. The only remaining approach, numerical simulation, is suitable for site and local scale for a long period of time. Despite the application of the hydrodynamic trapping mechanism in various time scales, it cannot be evaluated at regional- and basin-scale estimation. To calculate the storage capacity based on compressibility concept, a more comprehensive equation has been addressed recently including flow simulation employing volumetric formulas and more reservoir parameters.

In coal bed capacity estimation, the Langmuir equation provides a simple and efficient relation for single layer low-pressure conditions. In the case of high pressure and high temperature, Bi Langmuir, extended Langmuir, Sips, Langmuir-Freundlich, Toth, UNILAN, two-dimensional

EOS, LRC (loading ratio correlation), Dubinin–Radushkevich (D-R), and Dubinin-Astakhov (D-A) are more suitable.

One of the most efficient means for reservoir modeling is the TOUGH2 simulator developed in Rio Vista reservoir and an extension of EOS7R and EWASG modules also has been proposed to simulate the gas and water flow called EOS7C. A fluid/property module of TOUGH2 called ECO$_2$N has been utilized for CO$_2$ modeling in saline aquifers. TOUGHREACT, a non-isothermal reactive geochemical transport code, was utilized to simulate the CO$_2$ disposal in deep aquifers by entering the reactive chemistry term into the TOUGH2 framework. MUFTE-UG simulator has been evaluated for CO$_2$ sequestration in various fields of application such as simulation, CO$_2$SINK, and CO$_2$TRAP. Another survey with consideration of reactive flow and transport in deep saline aquifers has been performed using the GEM simulator. ECLIPSE and DuMux simulators are also taken into consideration in a study to understand the thermal effect during CO$_2$ injection and movement in the porous medium.

Six simulators including GEM, ECLIPSE, COMET2, SIMED II, GCOMP, and METSIM2 have been compared for CGS in coalbeds. GEM and SIMED II simulators are suitable for multi-component liquids while ECLIPSE and COMET2 can handle only two component fluids. Other comparison studies including TOUGHREACT, MUFTE, GEM, ECLIPSE, DuMux, COORES, FEHM, ROCKFLOW, SUTRA, and other types of simulators have been carried out throughout the world. Selecting the best simulator among those presented is highly based on the desired application. The ELSA simulator can be applied efficiently in semi-analytical estimation of fluid distributions. ROCKFLOW is suitable in the case of multi-phase flow and solute transport modeling. GEM is an aqueous geochemistry tool, while for the low temperature situation PHREEQC is more applicable. For multi-component, three phase, and 3D fluid flow simulation with consideration of reservoir heterogeneities, COORES would be a robust means. The new parallel multi-phase fluid flow simulator for CGS in saline aquifers called TOUGH+CO$_2$ has been developed on the basis of a modified TOUGH2 family of cods, TOUGH+ and TOUGH2-MP including all the ECO$_2$N feature capabilities and has proved to be a successful and robust means in a number of large scale simulation projects.

The CSEM have considered landing base imaging and passive magnetotelluric in deep crustal scale surveys in 2007. The 4D gravity and seismic methods have performed well in the Sleipner project. The 4D vertical seismic profiling (VSP) has been commonly used for quantitative monitoring of the CO$_2$ plume with tracer injection, well logging, and micro-seismic and pressure-temperature measurements with successful application at Frio and Nagaoka. In Frio Brine and Otway Pilot projects, tracer monitoring has been employed to assess the CO$_2$ breakthrough. The Eddy covariance and hyperspectral imaging in a shallow subsurface site are important computational issues that were examined to monitor the CO$_2$ leakage in Montana. Another successful surface monitoring technique tested at In Salah project was InSAR which incorporated to other monitoring techniques such as seismic, gravity, and electromagnetic. At Ketzin sequestration site, the monitoring methods included cross-hole resistivity, seismic, and microbiology with temperature and pressure profiling.

Author details

Kakouei Aliakbar[1], Vatani Ali[1*], Rasaei Mohammadreza[1] and Azin Reza[2]

*Address all correspondence to: avatani@ut.ac.ir

1 Chemical Engineering Department, College of Engineering, University of Tehran, Tehran, Iran

2 Department of Petroleum Engineering, Faculty of Petroleum, Gas and Petrochemical Engineering, Persian Gulf University, Bushehr, Iran

References

[1] IPCC: Special report on carbon dioxide capture and storage. 2005; Intergovernmental Panel on Climate Change (IPCC), Prepared by the IPCC work group III, Metz, B., Davidson, O., de Conick, H.C., Loos, M., Meyer, L.A. Cambridge University Press, Cambridge: 442.

[2] IPCC: Summary for policymakers. 2007; Intergovernmental Panel on Climate Change (IPCC), In: Solomon, S. (Ed.), Climate Change 2007: The Physical Sceince Basis. Contribution of working group I to the forth assessment report of the IPCC, Cambridge, United Kingdom and New York, NY, USA.

[3] Mann M., Bradley R.S., et al.: Global-scale temprature patterns and climate forcing over the past six centuries. 1998; Nature 392: 779–787.

[4] EIA: Energy-related carbon dioxide emission. 2006; Energy Information Administration (EIA)/DOE.International energy outlook.

[5] Tans P.: 2007. Retrieved from NOAA/ESRI: www.esrl.noaa.gov/gmd/ccgg/trends.

[6] Holloway S., Pearce J., et al.: Natural emissions of CO_2 from the geosphere and their bearing on the geological storage of carbon dioxide. 2007; Energy Conversion and Management 32: 1194–1201.

[7] Metz N.: Contribution of passenger cars and trucks. 2001; Environmental Sustainability Conference and Exhibition, Austria Graz.

[8] Bachu S.: Sequestration of CO_2 in geological media: criteria and approach for site selection in response to climate change. 2000; Energy Conversion and Management 41: 953–970.

[9] Bachu S. and Shaw J.C.: CO_2 storage in oil and gas reservoirs in western Canada: effect of aquifers, potential for CO_2-flood enhanced oil recovery and practical capacity.

In: Proceedings of the Seventh International Conference on Greenhouse Gas Control Technologies. 2005; Elsevier 1: 361–369.

[10] Preston B.L. and Jones R.N.: Climate change impacts on Australia and the benefits of early action to reduce global greenhouse gas emissions. 2006; retrieved from http:// csiro.au/files/p6fy.pdf.

[11] PhungQ.H., Kyuro S., et al.: Numerical simulation of CO_2 enhanced coal bed methane recovery for a Vietnamese coal seam. 2010; Journal of Novel Carbon Resource Science 2: 1–7.

[12] Khoo H.H. and Tan R.B.H.: Life cycle investigation on CO_2 recovery and sequestration. 2006; Environmental Scienceand Technology 40: 4016–4024.

[13] Allen D.J. and Brent G.F.: Sequestering CO_2 by mineral carbonation: stability against acid rain exposure. 2010; Environmental Science Technology 44(7): 2735–2739.

[14] Celia M.A. and NordbottenJ.M.: Practical modeling approaches for geological storage of carbon dioxide. 2009; Ground Water 47: 627–638.

[15] van der Zwaan B. and Smekens K.: CO_2 capture and storage with leakage in an energy-climate model. 2009; Environmental Modeling & Assessment 14: 135–148.

[16] Yang F., Bai B., et al.: Characteristics of CO_2 sequestration in saline aquifers. 2010; Petroleum Science 7(1): 83–92.

[17] PiresJ.C.M., Martins F.G., et al.: Recent developments on carbon capture and storage: An overview. 2011; Chemical Engineering Research and Design 89: 1446–1460.

[18] ZEP. Strategic overview.2007; European Technology Platform for Zero Emission Fossil Fuel Power Plants (ZEP).

[19] CamaraG.A.B., Andrade J.C.S., et al.: Regulatory framework for geological storage of CO_2 in Brazil-analyses and proposal. 2011; International Journal of Greenhouse Gas Control 5(4): 966–974.

[20] DOE/NETL: Carbon dioxide capture and storage RD&D roadmap. 2010 US Department of Energy.

[21] Spencer K.L., Bradshaw J., et al.: Regional storage capacity estimates: Prospectivity not statistics. 2010; CO_2 geological Storage Solution (CGSS), Australia.

[22] Holloway S. and Van Der Straaten R.: The joul II project-The underground disposal of carbon dioxide. 1995; Energy Conversion and Management 36(6–9): 519–522.

[23] Doughty C., Pruess K., et al.: Capacity investigation of brine-bearing sands of the Frio-formation for geological sequestration of CO_2. In: Proceedings of First National Conference on Carbon Sequestration. 2001; U.S. Department of Energy.

[24] Brennan S.T. and BurrussR.C.: Specific sequstration volumes: a useful tool for CO2 storage capacity assessment.2003; Open-File Report, U.S. Geological Survey, Second Annual Conference on Carbon Sequestration, Alexandria, Virginia, U.S. 03-452.

[25] Newlands I.K., Langford R.P., et al.: Assessing the CO2 storage prospectivity of developing economies in APEC applying methodologies developed in GEODISC to selected sedimentary basins in the Eastern Asian region. In: Proceedings of the Eigth International Conference on Greenhouse Gas Control Technologies. 2006; Elsevier.

[26] Bachu S., Bonijoly D., et al.: CO2 storage capacity estimation: methodology and gaps. 2007; International Journal of Greenhouse Gas Control 1(4): 430–443.

[27] USDOE: Carbon sequestration atlas of the United States and Canada. 2007; U.S. Department of Energy/NETL: 88.

[28] Bradshaw J., Bachu S., et al.: CO2 storage capacity estimation: issues and development of standards. 2007; International Journal of Greenhouse Gas Control 1(1): 62–68.

[29] Kopp A., Class H., et al.: Investigation on CO2 storage capacity in saline aquifers-Part 2: Estimation of storage capacity coefficient. 2009; International Journal of Greenhouse Gas Control, 3: 277–287.

[30] CSLF: A taskforce for review and development of standards with regards to storage capacity measurment; 2005.

[31] McCabe P.J.: Energy resources: cornucopia or empty barrel. 1988; AAPG Bull 82: 2110–2134.

[32] Bradshaw J., Bachu S., et al.: Discussion paper on CO2 storage capacity estimation (Phase 1): A taskforce for review and development of standards with regards to storage capacity measurment. 2005; CSLF-T 15.

[33] Kopp A., Class H., et al.: Investigations on CO2 storage capacity in saline aquifers: part 1. Dimentional analysis of flow processes and reservoir characteristics. 2009; International Journal of Greenhouse Gas Control 3(3): 263–276.

[34] FraileyS.M., Finley R.J., et al.: CO2 sequestration: storage capacity guidline needed. 2006; Oil & Gas Journal 104(30): 44–49.

[35] CO2CRC. Report: Storage capacity estimation, site selection and characterisation for CO2 storage projects, Edited by Kaldi JG., Gibson-Poole CM. 2008; RPT08-1001.

[36] Bachu S.: Comparison between methodologies recommended for estimation of CO2 storage capacity in geological media by the USDOE capacity and fairway subgroup of the regional carbon sequestration partnerships program (Phase III). 2008; The CSLFTask Force on CO2 Storage Capacity Estimation 04.

[37] Stevens S.H., Kuuskara V.A., et al.: Sequestration of CO2 in depleted oil and gas fields: global capacity, costs and barriers. In: Proceedings of Fifth International Con-

ference on Greenhouse Gas Control Technologies, CSIRO Publishing, Collingwood, Australia. 2001; 278-283.

[38] Doughty C., PruessK.: Modeling supercritical carbon dioxide injection in heterogeneous porous media. 2004; Vadose Zone Journal 3(3): 837–847.

[39] Taber J.J., Martin F.D., et al.: EOR screening criteria revisited-Part1: introduction to screening criteria and enhanced recovery field projects. 1997; SPEReservoir Engineering 12(3): 189–198.

[40] KovscekA.R.: Screening criteria for CO2 storage in oil reservoirs. 2002; Petroleum Science Technologies 20(7/8): 841–866.

[41] Bachu S. and Adams J.J.: Sequestration of CO2 in geological media in response to climate change: Capacity of deep saline aquifers to sequster CO2 in solution. 2003; Energy Conversion and Management 44(20): 3151–3175.

[42] Michael K., G. A., Shulakova V., Ennis-King J., Allinson G., Sharma S., Aiken T.: Geological storage of CO2 in saline aquifers-A review of the experience from existing storage operations. 2010; International Journal of Greenhouse Gas Control 4: 659–667.

[43] Juanes R., SpiteriE.J., et al.: Impact of relative permeability hysteresis on geological CO2 storage. 2006; Water Resource Reservoir 42(W12418): doi: 10.1029/2005WR004806.

[44] Lindeberg E., W.-B. D.: Vertical convection in an aquifer column under a gas cap of CO2. 1997; Energy Conversion and Management 38S: 229–234.

[45] Ennis-King J.P. and Paterson L.: Role of convective mixing in the long-term storage of carbon dioxide in deep saline formations. 2003; SPE 10: 349–356.

[46] Xu T., Apps J.A., et al.: Reactive geochemical transport simulation to study mineral trapping for CO2 disposal in deep arenaceous formations. 2003; Journal of Geophysical Research 108(B2): 2071–2084.

[47] Perkins E., C.-L. I., Azaroual M., Durst P.: Long term predictions of CO2 storage by mineral and solubility trapping in the WeyburnMidalereservoir. In: Proceedings of Fifth International Conference on Greenhouse Gas Control Technologies. 2004; Elsevier II2093–2101.

[48] Bachu S., Gunter W.D., et al.: Aquifer disposal of CO2: hydrodynamic and mineral trapping. 1994; Energy Conversion and Management 35(4): 269–279.

[49] Pingping S., Xinwei L., et al.: Methodology for estimation of CO2 storage capacity in reservoirs. 2009; Petroleum Exploration and Development 36(2): 216–220.

[50] De Silva P.N.K. and RanjithP.G.: A study of methodologies for CO2 storage capacity estimation of saline aquifers. 2012; Fuel 93: 13–27.

[51] DOE. Carbon sequestration atlas of the United States and Canada-II.2008; 1–142.

[52] Ehlig-Economides C. and Economides M.J.: Sequestering carbon dioxide in a closed underground volume. 2010; Journal of Petroleum Science and Engineering 70(1–2): 123–130.

[53] van der Meer L.G.H. and Yavuz F.: CO2 storage capacity calculations for the Dutch subsurface. 2009; Energy Procedia 1(1): 2615–2622.

[54] EcclesJ.K., Pratson L., et al.: Physical and economic potential of geological CO2 storage in saline aquifers. 2009; Environmental Science and Technology 43(6): 1962–1969.

[55] Ghanbari S., Al-Zaabi Y., et al.: Simulation of CO2 storage in saline aquifers. 2006; Chemical Engineering Research and Design 84(9): 764–775.

[56] Yang F., Bai B., et al.: Characteristics of CO2 sequestration in saline aquifers. 2010; Petroleum Science7(1): 83–92.

[57] OkwenR.T., TstewartM.T., et al.: Analytical solution for estimating storage efficiency of geologic sequestration of CO2. 2010; International Journal of Greenhouse Gas Control 4(1): 102–107.

[58] Zheng Z., Larson E.D., et al.: Near-term mega-scale CO2 capture and storage demonstration opportunities in China. 2010; Energy Environmental Science 3(9): 1153.

[59] White C.: Sequestration of carbon dioxide in coal with enhanced coalbedmethan recovery-a review. 2005; Energy Fuels 19(3): 659–724.

[60] Mazumder S. and Wolf K.H.: Differntial swelling and permeability change of coal in response to CO2 injection for ECBM. 2008; International Journal of Coal Geology 74(2): 123–138.

[61] van Bergen F., PagnierH.J.M., et al.: CO2-sequestration in the Netherlands: inventory of the potential for the combination of subsurface carbon dioxide disposal with enhanced coalbed methane production. In: Proceedings of Fifth International Conference on Greenhouse Gas Control Technologies, CSIRO Publishing, Collingwood, Australia. 2001; 555-560.

[62] Bachu S., Bonijoly D., et al.: Screening and ranking of sedimentary basins for sequestration of CO2 in geological media in response to climate change. 2003; Environmental Geology 44: 277–289.

[63] De Silva P.N.K., RanjithP.G., et al.: A study of methodologies for CO2 storage capacity estimation of coal. 2012; Fuel 31: 1–15.

[64] Rowley H.H. and Innes W.B.: Relationship between the spreading pressure, adsorption and wetting. 1942.

[65] Pan Z. and ConnelL.D.: A theoretical model for gas adsorption-induced coal swelling. 2007; International Journal of Coal Geology 69(4): 243–252.

[66] Saghafi A., Faiz M., et al.: CO2 storage and gas diffusivity properties of coals from Sydney Basin, Australia. 2007; International Journal of Coal Geology 70(1–3): 240–254.

[67] OzgenKaracan C. and Okandan E.: Assessment of energetic heterogenitiesof coals for gas adsorption and its effect on mixture predictions for coalbed methane studies. 2000; Fuel 79(15): 1963–1974.

[68] RyuY.K., Lee H.J., et al.: Adsorption equilibria of tuluene and gasoline vapors on activated carbon. 2002; Journal of Chemical Enginnering Data 47(5): 1222–1225.

[69] Ming L., Anzhong G., et al.: Determination of adsorbate density from supercriticl gas adsorption equilibria data. 2003; Carbon 41(3): 585–588.

[70] Ozdemir E., MorsiB.I., et al.: CO2 adsorption capacity of argonne premium coals. 2004; Fuel 83(7–8): 1085–1094.

[71] Siemons N. and Busch A.: Measurment and interpretation of supercritical CO2 sorption on various coals. 2007; International Journal of Coal Geology 69(4): 229–242.

[72] Day S., Duffy G., et al.: Effect of coal properties on CO2 sorption capacity under supercritical conditions. 2008; International Journal of Greenhouse Gas Control 2(3): 342–352.

[73] Dutta P., Harpalani S., et al.: Modeling of CO2 sorption on coals. 2008; Fuel 87(10–11): 2023–2036.

[74] Himeno S., Komatsu T., et al.: High-pressure adsorption equilibria of methane and carbon dioxide on several activated carbones. 2005; Journal of Chemical Enginnering Data 50(2): 369–376.

[75] BaeJ.S. and Bhatia S.K.: High-pressure adsorption of methane and carbon dioxide on coal. 2006; Energy Fuels 20(6): 2599–2607.

[76] PagnierH.J.M., et al.: CO2-sequestration in the Netherlands: inventory of the potential for the combination of subsurface carbon dioxide disposal with enhanced coalbed methane production. In: Proceedings of Fifth International Conference on Greenhouse Gas Control Technologies, CSIRO Publishing, Collingwood, Australia. 2001; 555-560.

[77] Vangkilde-pedersen T., AnthonsenK.L., et al.: Assessment European capacity for geological storage of carbon dioxide -the EU GeoCapacity project. 2009; Energy Procedia 1(1): 2663.

[78] Palarski J. and Lutynski M.: Capacity of an abandoned coal mibe converted into high pressure CO2 reserovir. Economic evaluation and risk analysis of mineral projects, London, UK. 2008; Taylor & Francis.

[79] Li D., Liu Q., et al.: High-pressure sorption isotherms and sorption kinetics of CH4 and CO2 on coals. 2010; Fuel 89(3): 569–580.

[80] NordbottenJ.M., Celia M.A., et al.: Injection and storage of CO2 in deep saline aquifers: analytical solution for CO2 plume evolution during injection. 2005; Transport Porous Media 58(3): 339–360.

[81] NordbottenJ.M., Celia M.A., et al.: Semianalytical solution for CO2 leakage through an abandoned well. 2005; Environmental Science Technology 39(2): 602–611.

[82] OmambiaA.N. and Li Y.: Numerical modeling of carbon dioxide sequestration in deep saline aquifers in Wangshang Oilfield-Jianghan Basin, China. 2010; Journal of American Science 6(8): 178–187.

[83] Zhang K., Moridis G., et al.: TOUGH+CO2: A multiphase fluid-flow simulator for CO2 geologic sequestration in saline aquifers. 2011; Computers &Geoscinces 37: 714–723.

[84] Oldenburg C., Pruess K., et al.: Process modeling of CO2 injection into natural gas reservoirs from carbon sequestration and enhanced gas recovery. 1995; Lawrence Berkeley National Laboratory Report, LBNL: 94720.

[85] Oldenburg C. and Pruess K.: EOS7R: Radionuclude transport for TOUGH2. 1995; Lawrence Berkeley National Laboratory Report, LBL 34868.

[86] Battistelli A., Calore C., et al.: The simulator TOUGH2/EWASG for modeling geothermal reservoirs with brines and non-condensible gases. 1997; Geothermics 26(4): 437–464.

[87] Pruess K., Oldenburg C., et al.: TOUGH2 user's guide, version 2.0. 1999; Ernest Orlando Lawrence Berkeley National Laboratory Report, LBNL 43134.

[88] Oldenburg C. and Pruess K.: EOS7C: Gas reservoir simulation for THOUGH2. 2000; Lawrence Berkeley National Laboratory Report, LBNL.

[89] OmambiaA.N. and Li Y.: Numerical modeling of carbon dioxide sequestration in deep saline aquifers in Wangshang Oilfield-Jianghan Basin, China. 2010; Journal of American Science 6(8): 178–187.

[90] Pruess K. and Spycher N.: ECO2N- A new TOUGH2 fluid property module for studies of CO2 storage in saline aquifers. 2006; Proceeding, TOUGH2 Symposium, Lawrence Berkeley National Laboratory Report, LBNL, California.

[91] Xu T., Apps J.A., et al.: Reactive geochemical transport simulation to study mineral trapping for CO2 disposal in deep arenaceous formations. 2003; Journal of Geophysical Research 108(B2): 2071–2084.

[92] Pruess K.: TOUGH2: A genaral simulator for multiphase fluid and heat flow. 1991; Lawrence Berkeley National Laboratory Report, LBNL, California: 29400.

[93] Xu T. and Pruess K.: Coupled modeling of non-isothermal multiphase flow, solute transport and reactive chemistry in porous and fractured media: 1. Model develop-

ment and validation. 1998; Lawrence Berkeley National Laboratory Report, LBNL, California: 42050.

[94] Xu T. and Pruess K.: On fluid flow and mineral alteration in fractured caprock of magmatic hydrothermal systems. 2001; Journal of Geophysics Reservoir 106: 2121–2138.

[95] Xu T., Apps J.A., et al.: Numerical simulation of CO2 disposal by mineral trapping in deep aquifers. 2004; Applied Geochemistry 19: 917–936.

[96] Hovorka S.D., Doughty C., et al. Testing efficiency of storage in the subsurface: Frio Brine pilot experiment (574).

[97] Helmig R., Class H., et al.: Architecture of the modular program system MUFTE-UG for simulating multiphase flow and transport processes in heterogeneous prous media. 1998; MathematischeGeologie2: 123–131.

[98] Ebigbo A., Bielinski A., et al.: Numerical modeling of CO2 sequestration with MUFTE-UG. 2006; Institute of Hydraulic Engineering, University of Stuttgart.

[99] Kempka T., Class H., et al.: Current status of the modeling activities at the Ketzin CO2 storage site. 2011; Geophysical Research Abstracts 13(EGU2011): 11591–11592.

[100] Pawar R.J., Zhang D., et al.: Preliminary geologic modeling and flow simulation study of CO2 sequestration in a depleted oil reservoir. NETL Carbon Sequestration Conference Proceedings.

[101] Kumar A., Noh M., et al.: Reservoir simulation of CO2 storage in deep saline aquifers. 2004; SPE/DOE Fourteenth Symposium on Improved Oil Recovery, Tulsa, USA(89343).

[102] Jiang X.: A review of physical modeling and numerical simulation of long-term geological storage of CO2. 2011; Journal of Applied Energy 88: 3557–3566.

[103] Law D H-S., van der Meer LGH., et al. Comparison of numerical simulators for greenhouse gas sequestration in coalbeds, Part III: More complex problems. NETL Carbon Sequestration Conference Proceedings.

[104] Class H., Ebigbo A., et al.: A benchmark study on problems related to CO2 storage in geologic formations. 2009; Computational Geosciences 13(4): 409–434.

[105] Parkhurst DL.andAppeloCAJ.: User's guide to PHREEQC (version 2)-A comuter program for speciation, batch-reaction, one-dimentional transport and inverse geochemical calculations. 1999; US Geological Survey Water-Resources Investigations Report: 99-4259.

[106] MillyPCD.: Moisture and heat transport in hysteretic, inhomogeneous porous media: anatric head based formulation and a numerical model. 1982; Water Resource Research 18(3): 489–498.

[107] Zhang K., W.Y.S., et al.: Parallel computing techniques for large-scale reservoir simulation of multi-component and multi-phase fluid flow. In: Proceeding of the 2001 SPE reservoir simulation synposium, Texas.2001; SPE.

[108] Wu Y.S., Zhang K., et al.: An efficient parallel-computing scheme for modeling nonisothermal multiphase flow and multicomponent transport in porous and fractured media. 2002; Advances in Water Resources 25: 243–261.

[109] Zhang K., Wu Y.S., et al.: Parallel commuting simulation of fluid flow in the unsaturated zone of Yucca Mountain, Nevada. 2003; Journal of Contaminant Hydrology.62–63.

[110] Zhang K., Wu Y.S., et al.: Flow focusing in unsaturated fracture networks: a numerical investigation. 2004; Vadose Zone Journal 3: 624–633.

[111] Zhang K., Doughty C., et al.: Efficient parallel simulation of CO2 geologic sequestration in saline aquifers. In: Proceeding of the 2007 SPE reservoir simulation synposium, Texas(106026). 2007; SPE.

[112] Yamamoto H., Zhang K., et al.: Numerical investigation concerning the impact of CO2 geologic storage on regional groundwater flow. 2009; International Journal of Greenhouse Gas Control 3(5): 586–599.

[113] Zhang K., Moridis G., et al.: TOUGH+CO2: A multiphase fluid-flow simulator for CO2 geologic sequestration in saline aquifers. 2011; Computers & Geosciences 37: 714–723.

[114] Azin R., JafariS.M., et al.: Measurement and modeling of CO2 diffusion coefficient in saline aquifer at reservoir conditions. 2013; Heat Mass Transfer 49: 1603–1612.

[115] Gholami Y., Azin R., et al.: Prediction of carbon dioxide dissolution in bulk water under isothermal pressure decay at different boundary conditions. 2015; Journal of Molecular Liquids 202: 23–33.

[116] Sheikha H., Pooladi-Darvish M., et al.: Development of graphical methods for estimating the diffusivity coefficient of gases in bitumen from pressure-decay data. 2005; Energy & Fuels, 19, 2041– 2049.

[117] JafariS.M., Azin R., et al.: Measurement of CO2 diffusivity in synthetic and saline aquifer solutions at reservoir conditions: the role of ion interactions. 2015; Heat Mass Transfer DOI 10.1007/s00231-015-1508-4.

[118] Azin R., Mahmudi M., et al.: Measurement and modeling of CO2 diffusion coefficient in saline aquifer at reservoir conditions. 2013; Central European Journal of Energy 3(4): 585–594.

[119] Zirrahi Z., Azin R., et al.: Prediction of water content of sour and acid gases. 2010; Fluid Phase Equilibria324: 80–93.

[120] Zirrahi Z., Azin R., et al.: Mutual solubility of CH4, CO2, H2S, and their mixtures in brine under subsurface disposal conditions. 2012; Fluid Phase Equilibria 299: 171–179.

[121] Zienckiewicz O., Taylor R., et al.: The Finite Element Method: Its Basis and Fundamentals. 2005; 1, Butterworth-Heinemann.

[122] Goel: Carbon capture and storage technology for sustainable energy. 2009; Jawaharlal Nehru University, New Delhi, India.

[123] Sharma S., Dodds K., et al.: Application of geophysical monitoring within the Otway Project S.E. Australia. Las Vegas 78th Annual SEG (Society of Exploration Geophysics) Conference. 2008; 2859–2863.

[124] Sreitch R., Becken L., et al.: Imaging of CO2 storage sites, geothermal reservoirs and gas shales using controlled-source magnetotellurics: Modeling studies, chewie der Ercle. 2010; Geochemistry 70(3): 63–75.

[125] Arts R., Eiken O., et al.: Monitoring of CO2 injected at Sleipner using time-lapsed seismic data. 2004; Energy Conversion and Management 29: 1383–1392.

[126] Arts R., Eiken O., et al.: Seismic monitoring at Sleipner underground CO2 storage site (North Sea). In: Geological Storage of Carbon Dioxide. 2004; Geological Society 233: 181–191.

[127] Chadwick A., Arts R., et al.: 4D seismic quantification of a growing CO2 plum at Sleipner, North Sea. In: Petroleum geology, North West Europe and Global perspectives-Proceedings of the 6th Petroleum Geology Conference. 2005; 15.

[128] Hovorka S.D. and Knox P.R.: Frio brine sequestration pilot in the Texas gulf coast. In: Proceedings of Sixth International Conference on Greenhouse Gas Control Technologies, Kyoto, Japan. 2003; 583–587.

[129] Freifield B., Trautz R., et al.: The U-tube; a novel system for acquiring borehole fluid samples from a deep geologic CO2 sequestration experiment. 2005; Journal of Geophysics Reservoir 110: B10203.

[130] Kikuta K., Hongo S., et al.: Field test of CO2 injection in Nagaoka, Japan. in: Proceedings of the 7th International Conference on Greenhouse Gas Control Technologies, Vancouver, Canada. 2005; 1367–1372.

[131] Zwingmann N., Mito S., et al.: Preinjectioncharacterisation and evaluation of CO2 sequestration potential in the Haisume formation, Niigata Basin, Japan-Geochemical modeling of the water-mineral-CO2 interation. 2005; Oil Gas Technology 60: 249–258.

[132] Hovorka S.D., Benson S.M., et al.: Measuring permanence of CO2 storage in saline formations: the Frio experient. 2006; Environmental Geoscience 13: 105–121.

[133] KharakaY.K., Cole D.R., et al.: Gas-water-rock interactions in Frio formation following CO2 injection: implication for the storage of greenhouse gase in sedimentary basins. 2006; Geology 34: 577–580.

[134] KharakaY.K., Cole D.R., et al.: Gas-water-rock interactions in sedimentary basins: CO2 sequestration in the Frio fromation, Texas, USA. 2006; Journal of Geochemical Exploration 89: 183–186.

[135] Mito S., Xue Z., et al.: Mineral trapping of CO2 at Nagaoka test site. In: Proceedings of Eighth International Conference on Greenhouse Gas Control Technologies, Trondheim, Norway. 2006.

[136] Saito H., Nobuoka D., et al.: Time-lapse crosswell seismic tomography for monitoring injected CO2 in an onshore aquifer, Nagaoka, Japan. 2006; Journal of Exploration Geophysics 37: 30–36.

[137] Xue Z., Tanase D., et al.: Estimation of CO2 saturation from time-lapse CO2 well logging in an onshore aquifer, Nagaoka, Japan. 2006; Journal of Exploration Geophysics 37: 19–29.

[138] Daley T.M., Solbau R.D., et al.: Continuous active-source seismic monitoring of CO2 injection in a brine aquifer. 2007; Geophysics 72: A57–A61.

[139] Muller N., RamakrishnanT.S., et al.: Time-lapse carbon dioxide monitoring with pulsed neutron logging. 2007; International Journal of Greenhouse Gas Control 1: 456–472.

[140] Daley T.M., Myer L., et al.: Time-lapse crosswell seismic and VSP monitoring of injected CO2 in a brine aquifer. 2008; Environmental Geology 54: 1657–1665.

[141] Doughty C., Freifield B., et al.: Site characterization of CO2 geological storage and vice versa: the Frio brine pilot, Texas, USA as a case study. 2008; Environmental Geology 54: 1635–1656.

[142] Mito S., Xue Z., et al.: Case study of geochemical reactions at the Nagaoka CO2 injection site, Japan. 2008; International Journal of Greenhouse Gas Control 2: 309–318.

[143] Onishi K., Ueyama T., et al.: Application of crosswell seismic tomography using difference analysis with data normalization to monitor CO2 flooding in an aquifer. 2009; International Journal of Greenhouse Gas Control 3: 311–321.

[144] Xue Z., Mito S., et al.: Case study: trapping mechanisms at the pilot-scale CO2 injection site, Nagaoka, Japan. 2009; Energy Procedia 1: 2057–2062.

[145] Freifield B., Trautz R., et al.: The U-tube; a novel system for acquiring borehole fluid samples from a deep geologic CO2 sequestration experiment. 2005; Journal of Geophysics Reservoir 110: B10203.

[146] Stalker L., Boreham C., et al.: Geochemical monitoring at the CO2CRC Otway project: tracer injection and reservoir fluid acquisition. 2009; Energy Procedia 1: 2119–2125.

[147] Keith C.G., RepaskyK.S., et al.: Monitoring effects of a controlled subsurface carbon dioxide release on vegetation using a hyperspectral image. 2009; International Journal of Greenhouse Gas Control 3: 626–632.

[148] LewickiJ.L., HilleyG.E., et al.: Eddy covariance observations of surface leakage during shallow subsurface CO2 releases. 2009; Journal of Geophysics Reservoir 114.

[149] RiddifordF.A., Tourqui A., et al.: A cleaner development: The In Salah gas project Algeria. In: Proceedings of the Sixth International Conference on Greenhouse Gas Control Technologies, Kyoto, Japan. 2003; 595–600.

[150] RiddifordF.A., Wright I.W., et al.: Monitoring geological storage in the Salah gas CO2 storage project. In: Proceedings of 7th International Conference on Greenhouse Gas Control Technologies, Vancouver, Canada. 2005; 1353–1359.

[151] Matheison A., Wright I.W., et al.: Satelite imaging to monitor CO2 movement at Krechba, Algeria. 2009; Energy Procedia 1: 2201–2209.

[152] Onuma T. and Ohkawa S.: Detection of surface deformation related to with CO2 injection by DInSAR at In Salah, Algeria. 2009; Energy Procedia 1: 2177–2184.

[153] Rutqvist J., Vasco D.W., et al.: Coupled reservoir-geochemical analysis of CO2 injection at In Salah, Algeria. 2009; Energy Procedia 1: 1847–1854.

[154] Forster A., Norden B., et al.: Baseline characterization of the CO2SINK geological storage site at Ketzin, Germany. 2006; Environmental Geosciences 13: 145–161.

[155] Juhlin C., Giese R., et al.: Case history: 3Dseismics at Ketzin, Germany: the CO2SINK project. 2007; Geophysics 72: B121–B132.

[156] Yordkayhun S., Julin C., et al.: Shallow velocity-depth model using first arrival traveltime inversion at the CO2SINK site, Ketzin, Germany. 2007; Journal of Applied Geophysics 63: 68–79.

[157] Kazemeini H., Juhlin C., et al.: Application of the continuous wavelet transform on seismic data for mapping of channel deposits and gas detection at the CO2SINK site, Ketzin, Germany. 2008; Geophysics Prospect 57: 111–123.

[158] Giese R., Henninges J., et al.: Monitoring at the CO2SINK site: a concept integrating geophysics, geochemistry and microbiology. 2009; Energy Procedia 1: 2251–2259.

[159] Prevedel B., Wohlgemuth L., et al.: The CO2SINK boreholes for geological CO2-storage testing. 2009; Energy Procedia 1: 2087–2094.

[160] Schilling F., Borm G., et al.: Status report on the first European on-shore CO2 storage site at Ketzin, Germany. 2009; Energy Procedia 1: 2029–2035.

4

Carbon Footprint as a Tool to Limit Greenhouse Gas Emissions

Francesco Fantozzi and Pietro Bartocci

Additional information is available at the end of the chapter

Abstract

The Carbon Footprint is the amount of greenhouse gases (GHG) produced during the life cycle of a product, a process, or a service (expressed in equivalent tons of carbon dioxide per functional unit of analyzed product/process/service). The patterns of fossil fuel combustion, carbon capture and sequestration, and conventional and unconventional fossil fuel production, but also the emissions linked with consumer behavior, can be analyzed considering their carbon footprint. In this chapter the carbon footprint tool is introduced, linking it to fossil energy systems and renewable energy systems, as well as the main products on the market, to provide information on which technology should be promoted to reduce GHG emissions.

Keywords: Carbon Footprint, ISO 14067, GHG, Life Cycle Assessment

1. Introduction

Carbon footprint, as an indicator of the impact of the emissions of GHG of products and services, is interesting for enterprises, consumers, and politicians [1]. Investors control the carbon footprint of their products as it is an indicator of their investment risk. Purchasing managers are interested in the carbon footprint of the goods that they are dealing with, and the market is beginning to offer consumers carbon-labeled products. These are the reasons for the popularity of product carbon footprint. It is defined as the mass of cumulated CO_2 emissions that can be measured through a supply chain or through the life cycle of a product [2]. The average per capita carbon footprint of continents and of the most important nations is reported in Table 1 (data are expressed in equivalent tons of carbon dioxide per capita per year). Also the contribution of different sectors is reported (expressed in percentage).

	Country footprint [tCO2e/p]	Construction (%)	Shelter (%)	Food (%)	Clothing (%)	Manufactured products (%)	Mobility (%)	Service (%)	Trade (%)
Europe	13	9	21	16	3	12	21	17	6
USA	29	7	25	8	3	12	21	16	8
Canada	20	8	18	8	2	9	30	18	6
South America	5	6	8	36	3	8	22	13	5
Russian Federation	10	9	40	15	1	3	16	17	1
Asia	7	11	14	24	4	11	19	16	4
Africa	2	6	13	40	2	6	10	22	3
Australia and New Zealand	16	8	18	18	3	9	19	16	13

Table 1. Carbon footprint of continents and most important nations [3]

Carbon footprint is most appropriately calculated using life-cycle assessment or input-output analysis [3,4]. In this sense it is based on the ISO 14040 [4] and ISO 14043 [5] norms, on life cycle assessment (LCA). Specific norms for carbon footprint of enterprises and products are ISO 14064 (part 1,2, and 3) [6-8], ISO 14067 [9], and PAS 2500 [10]. Carbon footprint calculation process is shown in Figure 1.

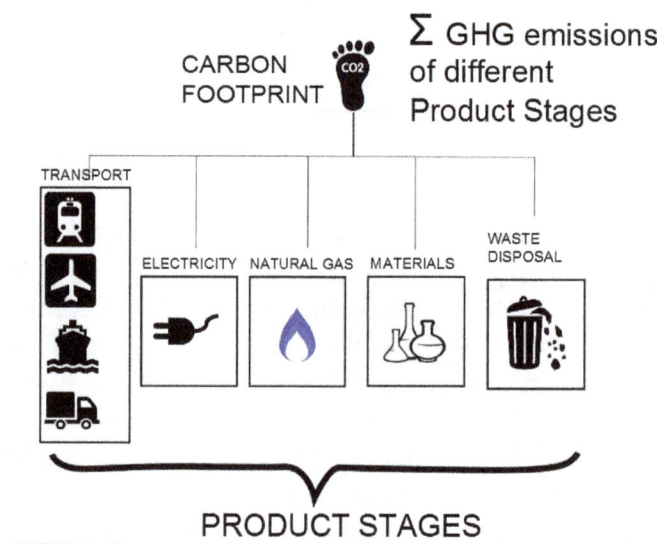

Figure 1. Carbon footprint calculation

Main emissions due to the most important processes are added through the whole life cycle. Carbon footprint of a purchased good or service can be calculated using Equation 1 [11].

$$PCF = S1 + S2 + S3 + S4 + OE \qquad (1)$$

Where:

- S1 is the sum across purchased goods and services;

- S2 is the sum of emissions due to material inputs;

- S3 is the sum of emissions due to transport of material inputs;

- S4 is the sum of emissions due to waste outputs;

- OE stands for other emissions emitted in provision of the good or service.

In order to calculate carbon footprint, it is very important to consider the boundaries of the process: which emissions should be considered in calculation of the footprint? This problem can be solved by considering three definitions: Scope 1, Scope 2, and Scope 3.

Scope 1 indicates direct emissions, for example, on-site emissions; Scope 2 indicates emissions embodied in the purchased energy; and Scope 3 indicates all the emissions not covered under Scope 2, such as those associated with transport of goods and waste disposal [12].

Another important aspect is the functional unit, which is defined as a measure of the function of the studied system, and it provides a reference to which the inputs and outputs can be related. This enables the comparison of two essential different systems.

2. Carbon footprint of renewable energy systems

2.1. Carbon footprint of transport fuels

The carbon footprint of transport fuels has been analyzed in several studies starting from 1990. One of the most important is the study realized by Sheehan et al. [13] at National Renewable Energy Laboratory of the United States.This is an LCA study that includes the impact of CO_2 emissions. Most important operations belonging to the petroleum diesel product system include crude oil extraction, its transport to an oil refinery, crude oil refining to diesel fuel, its transportation to the user, and its use in a bus engine.

In addition to energy and environmental outputs in each step, energy and environmental inputs from raw materials use are also included. Generally, life cycle flows include all raw materials used for extraction. Likewise, life cycle flows from intermediate energy sources such as electricity, back to the extraction of coal, oil, natural gas, limestone, and other primary resources should be included.

Life cycle presents a typical allocation case because the refining process is a multiple product process and the other sub-products obtained during diesel production are shown in Figure 2, together with the definition of the most important processes involved in the refining step.

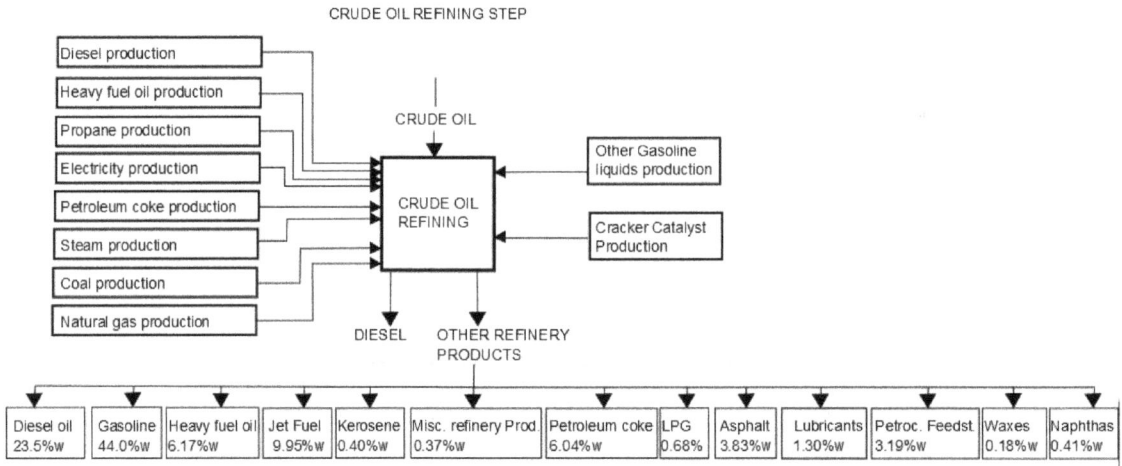

CRUDE OIL REFINING STEP

Diesel production
Heavy fuel oil production
Propane production   CRUDE OIL
Electricity production
Petroleum coke production   Other Gasoline
                                                         liquids production
Steam production   CRUDE OIL
                                         REFINING
Coal production   Cracker Catalyst
                                                         Production
Natural gas production

DIESEL   OTHER REFINERY
              PRODUCTS

Diesel oil	Gasoline	Heavy fuel oil	Jet Fuel	Kerosene	Misc. refinery Prod.	Petroleum coke	LPG	Asphalt	Lubricants	Petroc. Feedst.	Waxes	Naphthas
23.5%w	44.0%w	6.17%w	9.95%w	0.40%w	0.37%w	6.04%w	0.68%	3.83%w	1.30%w	3.19%w	0.18%w	0.41%w

Figure 2. Crude oil refining step [13]

The final results show that diesel production and use account for a total emission of CO_2 of 633 $gCO_2eq/bhp-h$. The processes that contribute most to the release of CO_2 emissions are refining (which is responsible for 10% of the emissions) and petroleum combustion in the engine (which is responsible for 87% of the emissions).

2.2. Carbon footprint of electricity generation through fossil fuels

The electricity supply sector is responsible for over 7,700 million tonnes of CO_2 emissions annually (2,100 Mt C/yr); being 37.5% of total CO_2 emissions [14]. The annual carbon emissions, associated with electricity generation, is projected to surpass the 4,000 Mt C level by 2020 [15]. Past and projected electricity production from fossil fuels is shown in Table 2 and also CO_2 emissions per kWh.

	1995	2000	2010	2020
Coal	4,949	5,758	7,795	10,296
Natural Gas	1,932	2,664	5,063	8,243
Oil	1,315	1,422	1,663	1,941
Average GHG emissions (g CO_2/kWh)	158	157	151	147

Table 2. Past and projected global production from the electricity generating sector (TWh/yr) and average CO_2 emissions per kWh [14]

The efficiencies of modern thermal power stations using the steam cycle can exceed 40% based on lower heating value, although the average efficiency of the installed stock worldwide is closer to 30%. Recently, efficiencies of 48.5% have been reported and, with further development, by 2020, they could reach 55% at costs only slightly higher than current technology.

Physical carbon sequestration is more useful with the emissions of large point sources of CO_2 such as power plants. It can be captured either before combustion, in an IGCC or in a reforming process (transforming steam to methane), or after combustion from the flue gas stream using amine solvents, for example. The volume percentage of CO_2 in exhaust flue gases is between 4% (for gas turbines) and 14% (for a pulverized coal-fired plant), which means that large volumes of gas have to be treated using efficient solvents, and this will result in high-energy consumption because of solvent regeneration. These techniques will achieve an efficiency of 80–90% in carbon capture. Other carbon capture techniques include cryogenics, membranes, and adsorption. After the CO_2 has been captured, it is pressurized, typically up to 100 bar, before transportation to storage areas. CO_2 capture and compression imply a decrease on the thermal efficiency of a power plant, which has been estimated to be equal to 8–13%. The cost of CO_2 capture in power plants comprises between \$30 and 50/t CO_2 of emissions; while the cost of CO_2 transportation is influenced by the distance and the capacity of the pipeline and ranges between \$1 and 3/t CO_2 per 100 km. The cost of underground storage, which excludes the costs due to compression and transport, is estimated to range between \$1 and 2/t CO_2 stored. With the development of new technologies, for example the development of new solvents and system components, the costs of carbon capture and storage would decrease.

2.3. Carbon footprint of residential heating systems based on fossil fuels

Glaeser and Kahn [16] evaluated the emissions released by American households for heating purposes. The two primary heating sources for households are fuel oil and natural gas. On the one hand in the United States, the use of fuel oil is pretty rare, with the exception of the Northeast, and it is used as a source of home heating in few metropolitan areas; on the other hand, natural gas is the most common home heating source; and in some areas electricity is also used. Natural gas consumption is driven primarily by climate.

For fuel oil and natural gas, there are conversion factors that enable to move from energy use to CO_2 emissions. In the case of fuel oil, the factor is 22.38 lb of CO_2 per gallon.

It can be considered that about 20,000 kWh/yr are required to heat a typical house in developed countries. If hard coal, oil, natural gas, and LPG are used, the annual total CO_2 emissions are 8,280 kg CO_2/yr, 6,280 kgCO_2/yr, 4,540 kg CO_2/yr, and 5,180 kg CO_2/yr, respectively [17]. These data agree with those reported by Johnson [18], which are shown in Figure 3.

2.4. Life cycle carbon footprint of shale gas

Recent advances in drilling and fracking technologies have made the access to huge deposits of natural gas in shale deposits technically and economically feasible. These are located across the United States and elsewhere [19,20], and thus shale gas production has grown about 48% per year from 2006 to 2010 in the United States. This fact will influence the American and the world energy outlooks for the near future, together with the variation in the oil price [21]. The

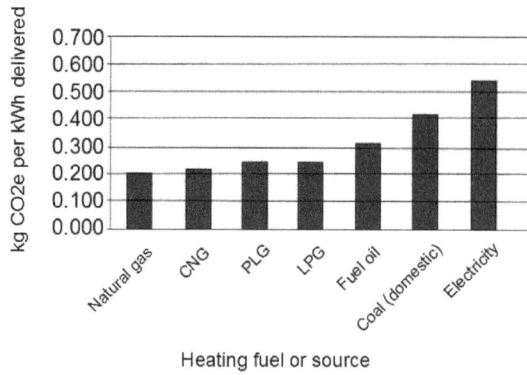

Figure 3. Carbon footprint of different heating fuels [18]

growth of the shale gas industry has brought important benefits, such as significant job growth, decoupling gas and oil prices, providing an alternative to the more polluting use of oil in transportation and of coal in power generation [22,23]. The carbon footprint of shale gas can be calculated evaluating or measuring the direct CO_2 emissions from its final use and evaluating indirect CO_2 emissions produced from fossil fuels used to extract, develop, and transport it. Also methane fugitive emissions and emissions from venting have to be considered. Literature studies have shown that the indirect CO_2 emissions throughout shale oil life cycle are relatively small than that of the direct combustion of the fuel. In fact indirect emissions range between 1 and 1.5 g CO_2/MJ−1 [24], whereas direct emissions range between 13-15 g CO_2/MJ [25,26]. Indirect emissions from shale gas are comparable with those due to conventional gas production [26].

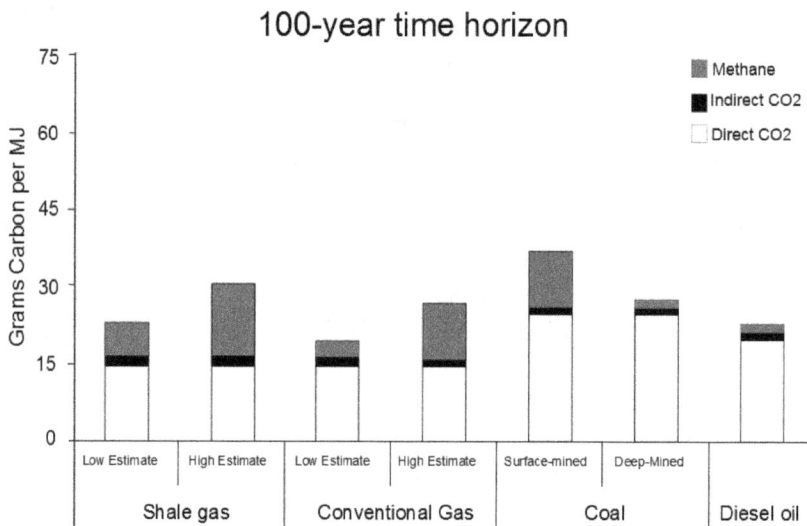

Figure 4. Comparison of GHG emissions from shale gas and conventional natural gas with low and high estimates of fugitive methane emissions, surface-mined coal, deep-mined coal, and diesel oil; time horizon equal to 100 years [27]

From the most important studies available in literature, it can be inferred that both carbon footprints of shale gas and of conventional gas are dominated by direct CO_2 emissions and fugitive methane emissions. In Figure 4 direct emissions of CO_2 during combustion of shale gas, conventional gas, coal, and diesel oil are represented with white bars, whereas indirect emissions occurring during the development and the use of the energy sources are represented in black bars, and fugitive emissions of methane are represented with grey bars. All the emissions have been normalized to the quantity of energy released during combustion.

3. Carbon footprint of renewable energy systems

3.1. Carbon footprint of transport biofuels (biodiesel, bioethanol, biomethane)

The GHG emissions released during biodiesel life cycle are about 40–65% of those released during conventional diesel life cycle. For bioethanol technologies, the GHG emissions are deeply influenced by the technology. Emissions of the whole life cycle of bioethanol produced from corn can be about 80–90% of those of competitor fossil fuels. For bioethanol produced from sugar cane, a reduction of 75-80% in fossil fuels emissions can be achieved. Important factors that influence the final results are the amounts and type of fossil fuels used in the life cycle as energy carriers to produce, transport, and process the feedstock. Also non-CO_2 emissions, generated during the cultivation phase, such as N_2O, have to be considered. Besides, the efficiency in the conversion process is important too, together with the degree to which biomass is used to provide the energy required by the process, and feedstock yields during the cultivation phase. The mass and energy balances are also influenced by the capacity of the bioenergy plant and the scale of the project. In the case of large-scale projects, there will be important land use changes that can influence carbon stocks in the soil. Table 3 shows GHG emissions per kilometer travelled.

Transportation Fuel	GHG Emissions (gCO_2-eq./km)
Bioethanol from sugar cane	50-75
Bioethanol from other crops (corn, sugar beet, wheat)	100-195
Biogas	25-100
Biodiesel (rapeseed, soy, sunflower)	80-140
Fischer Tropsch diesel from biomass	15-55
Bioethanol from lignocellulose	25-50
Gasoline	210-220
Diesel	185-220
Natural gas	155-185

Table 3. GHG emissions per kilometer travelled using renewable fuels and fossil fuels [28]

3.2. Carbon footprint of electricity generation through renewable energy

The carbon footprint of electricity generation through RES (Renewable Energy Systems) are described in this section. Hydro-electricity is described first, and then wind power, followed by bioenergy systems and solar energy.

Hydro-electricity is the most developed renewable resource worldwide, even if it has to face social and environmental barriers [29]. In fact societal preferences are difficult to predict, while hydro-sites are often difficult to reach, which results in high transmission and capital investment costs. These are difficult to be accepted by private power companies. The global economic hydropower potential ranges between 7000 and 9000 TWh per year. Particularly rural communities without electricity appear to be convenient for small (<10 MWe), mini- (<1 MWe), and micro- (<100 kWe) scale hydro schemes. They have low environmental impacts, and generation costs are around 6–12 c/kWh. Emissions of GHG linked with hydro-electricity operation are due to flooding of land upstream of a dam that can imply a loss of biological carbon stocks and can produce methane emissions due to vegetation decomposition.

Wind power is a technology that has been developed recently. It has an intermittent flow and produces about 4% of total global electricity. In 2013 the production capacity reached 282,000 MWe, which implies a huge development, respect from year 2000 [30]. Denmark is producing about 40% of its total electricity consumption from wind power, and it's one of the main exporters of wind turbine technology. Many wind turbines will be sited off-shore. On the one hand in this way future demand can be met, the advantage is to increase the rated output to more than 3 MWe, to decrease costs linked with operation and maintenance, to have more reliable plants; on the other hand the cost of investment for wind turbines is decreasing, while the installed capacity increases. So wind power is becoming competitive with other sources of energy in highly windy areas. The costs of electricity generation in this case range between 3 and 5 c/kWh. The investment costs will fall from $1000 to $635/kWe and operating costs will decrease to about 0.01 c/kWh - 0.005 c/kWh [31].

Bioenergy can be produced from agricultural and forestry residues, animal effluents, the organic fraction of municipal solid wastes, and dedicated energy crops. Since biomass is widely spread in the territory, it is an interesting source of energy for rural and mountain areas. The challenge is to optimize the production of biomass, collection and logistics, optimize its conversion to energy and delivery to the end user, to provide a service that is economically competitive with that obtainable using other fossil fuels. Residual biomasses, such as bagasse, rice husks, straw, olive husk, bark, and sawdust often have a corresponding cost for disposal. Therefore, biomass-to-energy conversion, in the case of residues, can have good economic performance, especially in rural areas, where there is abundance of them. Denmark produces about 40% of the electricity it consumes through cogeneration plants, using wood waste and straw. Also biogas is produced from animal breeding effluents. Energy crops are less promising in the short term due to their higher production costs in terms of $/GJ of available energy. Also the competition for land use with food crops is becoming an issue. Biomass as a fuel is more reactive than coal if it is used in gasification process, which promotes the use of biomass in IGCC systems, that are approaching to commercial realization. Besides if coupled with carbon capture, biomass integrated gasification combined cycle can be a carbon-negative technology, because CO_2 is absorbed during biomass cultivation and production, and it is not released in the IGCC plant, due to carbon capture. On the one hand, capital investment for a biomass gasification–combined cycle plant, working with an high pressure reactor, is decreasing from \$2000/kWe to \$1100/kWe by 2020; on the other hand operating costs (fuel supply included) will decline from 3.98 to 3.12 c/kWh [31]. Actually operation costs for a traditional plat working with boiler plus steam turbine are about 5.50 c/kWh.

The cost of solar photovoltaic (PV) is slowly decreasing from \$5,000/kWe to \$4,000/kWe installed. The increase in the installed capacity corresponds to an increase in scale-up of manufacturing plants and the use of mass production techniques that are the main reasons for costs reduction. Also operating costs are quite high, being about 20–40 c/kWh. Promising new applications for solar PV are represented by grid connected buildings and by large installations (up to 1 MWe), which are pushing innovation in inverters and net metering systems. Other important markets for photovoltaic power systems are off-grid applications for rural areas, especially in developing countries where there is a need for electrification projects. The worldwide installed PV capacity is estimated to be about 178 GWe in 2014, while it will reach about 400 GWe in 2020. Conversion efficiencies of silicon cells are continuously improving. The efficiency of commercial monocrystalline modules is about 13–17%, whereas the efficiency of multicrystalline module is about 12–14%. Literature studies show that a single factory of 400 MWe capacity (obtainable with 5 million panels) can reduce production costs of 75%, due to economies of scale [32]. Neij [33] calculated that a \$100 billion investment in manufacturing capacity would be needed in order to reach an acceptable generating level of 5 c/kWh (excluding back-up supply or storage costs). Capital costs for concentrated solar will fall from \$4000/kWe to \$2500/kWe by 2030 (Table 4) [34].

Technology	PF + fgd, NO$_x$, etc.	IGCC and super-critical	CCGT	PF + fgd + CO$_2$ capture	CCGT +CO$_2$ capture	Nuclear	Hydro	Wind Turbines	Biomass IGCC	PV and Solar thermal
Energy source	Coal	Coal	Gas	Coal	Gas	Uranium	Water	Wind	Biofuel	Solar
Emissions (gC/kWh)	229	190-198	103-122	40	17	0	0	0	0	0
Reduction potential to 2020 (MtC/yr)	Baseline	55	103	5-50	N.A.	191	37	128	77	20
Cost of C reduction (\$/t C avoided)	Baseline	-10-40	0-156	159	71-165	-38-135	-31-127	-82-135	-92-117	175-1400

PF, pulverised fuel; fgd, flue gas desulphurization; IGCC, integrated gasification combined cycle

Table 4. Cost estimates of alternative mitigation technologies in the power generation sector compared to baseline pulverized coal-fired power plant and natural gas Combined Cycle with Gas Turbine (CCGT) power stations and the potential reductions in CO_2 emissions to 2020 [14]

3.3. Carbon footprint of residential heating systems based on renewable energy

Heat production and hot water supply to buildings are essential and important worldwide. The problem is how to produce them in a sustainable way, replacing fossil fuels. Today, it is intensively being discussed how to do so in the best way in future energy systems in which the combustion of fossil fuel should be reduced or completely avoided. One way could be through the promotion of low energy buildings in which the consumption of energy can be reduced or even removed (through the use for example of solar thermal heating systems). Another way could be the one to use excess heat produced from the industrial sector, waste incineration, power stations based on large-scale exploitation of geothermal energy, solar thermal energy, and heat pumps powered by excess wind energy. In these cases, a district heating network becomes essential. Table 5 shows the comparison of GHG emissions for different household.

Heat Source	GHG Emissions (gCO$_2$-eq./MJ)
Biomass (i.e., wood chips, pellets)	520
Geothermal	15
Solar thermal	1030
Coal	110150
Oil	90120
Natural gas	7085
Electricity from natural gas (space heating)	180210
Electricity from oil (space heating)	265295
Electricity from coal (space heating)	290320

Table 5. GHG emissions per unit output in the heating sector (taken from [28])

The development of district heating systems is linked with the development of other systems such as combined heat and power systems, which generate waste heat, together with power. These increase the fuel use efficiency [35]. Also heat pumps should be introduced in residential heating systems [18]. In some countries like Norway, district heating system's GHG emissions have been compared with those of individual heating systems and it has been found that the first have lower CO$_2$ emissions.

4. Carbon footprint of products

4.1. Carbon footprint in the food industry

Food industry sector is one of the major contributors to climate change [36]. In Sweden, it has been estimated that about 25% of GHG emissions from the private sector are due to the consumption of food [37]. In the European Union, food industry contribution to GHG emissions is estimated to be about 31% [36,38]. GHG emissions in the transport sector are mainly due to CO_2; while in agriculture most emitted GHG are methane (CH_4) and nitrous oxide (N_2O). The CO_2 emitted from land use change represents also an important source of emissions of the food production system. Starting from the publication of the Fourth Assessment Report of the IPCC in 2007 [39], the calculation of food carbon footprint has become more and more popular. Food carbon footprint is calculated by companies also for marketing purposes [40-42]. Also research efforts in the calculation of carbon footprint of food and in the estimate of its uncertainty have increased in recent years [43-46] (see Figure 5). Challenges in calculating the carbon footprint of food products can be linked with the functional unit, system boundaries and allocation, land use change, carbon sequestration in soils, uncertainties, and variation.

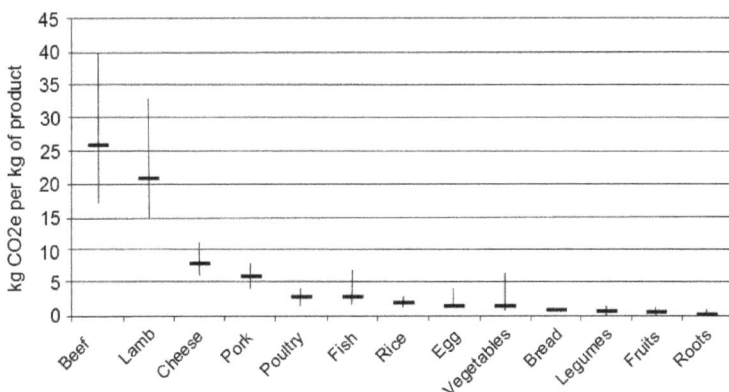

Figure 5. Carbon footprint of different types of food products at retail. Average values estimated to be representative for food products sold on the Swedish market. Error bars show ranges of values found in the literature. Emissions from land use change and carbon stock changes in soils are not included [47]

Besides marketing purposes, carbon footprint is calculated in the food sector also with the aim at reducing its value, producing more sustainable food. The main ways to reduce the product carbon footprint of food are as follows: by reducing emissions of CO_2 due to energy use in agriculture (for example improving energy efficiency and using renewable energy sources) and reducing CH_4 emissions from enteric fermentation and N_2O emissions from fertilizer nitrification in soil. CH_4 emissions can be reduced to some extent by altering the diet fed to ruminants [48], but the risk of pollution swapping is great [49,50]. N_2O emissions from soils can be reduced by optimizing nitrogen use and promoting N_2O inhibitors.

Another way to reduce GHG emissions in the food sector is changing the consumption patterns [51-54]; for example, switching from diets based on meat to diets in which proteins are also supplied by vegetables.

Being on-farm emissions (from cultivation and animals breeding) the most important source of GHG in food life cycle, numerous studies have tried to reduce them. Ahlgren [55] has used LCA to evaluate the use of biofuels in tractors and the substitution of mineral nitrogen fertilizers. This implied that 3–6% of a farm's available land was needed to produce the required biomass (to produce biofuels and fertilizer).

Another issue is represented by dairy production and the carbon footprint of milk [56,50]. An important area of research is the production of animal feed for the different diets used in livestock production [57-58].

4.2. Carbon footprint in the textile industry

Many enterprises in the textile and clothing industry are involved in product carbon footprint calculation. They range from fiber manufacturers (e.g., Lenzing, Advansa, Dupont) to producers of flooring material (e.g., InterfaceFlor, Desso, Heugaveld), to fashion brands (united in the Sustainable Apparel Coalition), to other organizations (European Commission and the Dutch branch organization Modint). They are using LCA to calculate the environmental impacts of textile-related products. Also educational textile and fashion institutes (e.g., the Amsterdam Fashion Institute) are promoting life cycle thinking, picking up the signals from companies and other organizations. A literature survey [59] shows that Collins and Aumônier [60] compiled a LCI (Life Cycle Inventory) on textile products upon references dating from 1978 to 1999. Another research executed by Kalliala and Talvenmaa [61] reports, for example, spinning energy, which is derived from a study out of 1997. In-depth investigation on weaving led to the research of Koç and Çinçik [62]. In the recent work of Shen [63], non-renewable energy use for the production processes of different fabrics is given, based upon a report from 1997 [64].

Walser et al. [65] have published a LCA study using inventory data for polyester (PET) textile production. The authors also noticed that the data in the Ecoinvent database [66] on cotton and bast fibers do not specify the yarn size, which has an important influence on energy use.

Figure 6 presents the carbon footprint of cotton textiles and of synthetic textiles. In the case of cotton, different yarn thickness are taken into account. They are expressed on decitex (abbreviated dtex). In the case of synthetic textiles, only yarn thickness of 70 dtex is taken into account.

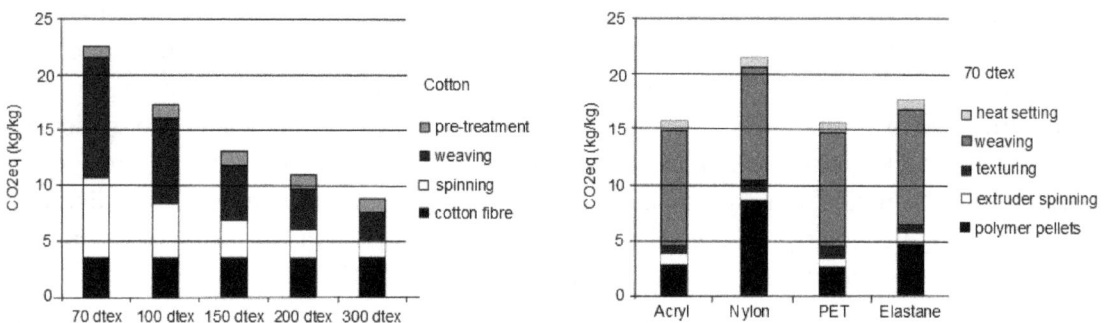

Figure 6. Carbon footprint of cotton textiles with yarn thickness comprised between 70 and 300 dtex (left) and synthetic textiles - acryl, nylon, PET, elastane-, with yarn thickness of 70 dtex (right) [59]

4.3. Carbon footprint in the cement industry

The cement industry is one of the sectors that contributes most to climate change, accounting for roughly 5% of the total CO_2 emissions worldwide [67]. Therefore reducing these emissions is a primary goal in order to comply with the objectives laid down in the Kyoto protocol to combat climate change. Currently, the cement industry, belonging to the WBSCD (World Business Council for Sustainable Development), has launched the Cement Sustainable Initiative program to meet the challenges of sustainable development. The carbon footprint is the most promising tool to evaluate the impact of carbon emissions of different products and can be an indicator to be used for eco-labeling. Several efforts have been made to develop it [68]. The study of Cagiao et al. [69] is based on the MC3 approach, also called organization-product-based-life-cycle assessment (OP-LCA). Given its top-down approach, this methodology first allows the organization's footprint to be calculated and then distributing it among the products that it manufactures. Some of the advantages are as follows:

a.   It is a single methodology to be used both for organizations and products.

b.   It uses all the financial accounts as input data.

c.   The information flows automatically through the value chain.

d.   The scope is always the same for all the analyses.

e.   It is simple and easy-to-understand and adaptable.

f.   Both the carbon footprint and the ecological footprint of the organization can be obtained [70-72].

The study of Cagiao et al. [69] was carried out with three potential scenarios in mind: case A pertaining to a conventional integral plant; case B which refers to a grinding plant; and case C, an integral plant which has been subject to the best available technical improvements. All the plants have the same productivity of 1,000,000 t/year. A summary of main results is proposed in Table 6.

Process	Emissions	Case
Carbon footprint of cement industry	1,003,555.2 tCO$_2$/year	Case A
	907,384.2 tCO2/year	Case B
	790,278.3 tCO2/year	Case C
Carbon footprint of one ton of cement	1.00 tCO2/tcement	Case A
	0.91 tCO2/tcement	Case B
	0.79 tCO2/tcement	Case C
Main parts of the carbon footprint produced	Direct emissions (75.33%) and wastes (17.98%)	Case A
	Materials (75.07%) and services (14.60%)	Case B
	Direct emissions (77.06%) and wastes (15.18%)	Case C
Reduction of total footprint by using BATs	213,276.9 tCO2/year (21.25% of initial carbon footprint)	

Table 6. Carbon footprint of cement production [69]

5. Product Carbon Footprint (PCF) case study

Fantozzi et al. [73] presents the study of the carbon footprint of a typical food product in Central Italy: truffle sauce. This is a mixture of vegetable oil and truffle in proportions of 33% and 67% respectively and minor components and spices (garlic, salt, pepper, etc.). Both truffles and olives are cultivated and harvested in a farm in Umbria (Italy). Olives are crushed in a mill that is situated few kilometers from the farm. Once it has been produced, the extra virgin oil, together with the truffle, is transported to another facility to produce bottled truffle sauce. The carbon footprint calculation is based on ISO 14076 technical standard. Product Category Rules (PCR) have been developed (see Table 7).

Stage	Rule	Description
Scope and functional unit	Scope	Calculate PCF of truffle sauce (expressed in kgCO$_2$eq/kg product)
	System boundary	Cultivation, transformation, packaging, and waste disposal are taken into account, while consumption is neglected
	Allocation	Allocation based on system expansion has to be preferred to allocation based on mass and economic value
Product definition	Truffle sauce	Truffle sauce is a mixture of vegetable oil and truffle in proportions of 33% and 67% respectively and minor components and spices (garlic, salt, pepper, etc.) that were not considered in the analysis
	LC stages	Cultivation; Milling; Truffle production; Transport; Sauce production
PCF calculation	Software	Simapro software was used to design process tree, and calculate PCF, based on the impact method GWP 100 years. Cut-off on processes impact is set to 1% to ease results view
	Data uncertainty	Data uncertainty was measured based on used instruments precision and on the uncertainty of Simapro datasets
Results communication	Label	A carbon footprint label was designed for the package

Table 7. Product Category Rules of truffle sauce [73]

The cut-off threshold on life cycle processes is about 1%. This decision is due to the need to simplify the process tree diagram. All the calculations are referred to the growing season 2011/2012.

The boundaries of the system analyzed are shown in Figure 7. Truffle sauce life cycle has been divided in the following product stages:

- cultivation;

- truffle production;

- truffle sauce production;

- packaging.

This is a clear example of a cradle-to-grave study, so GHG fluxes comprise also disposal of the packaging. The consumption phase is not considered in the study. The functional unit is 1 kg of truffle sauce.

TRUFFLE SAUCE

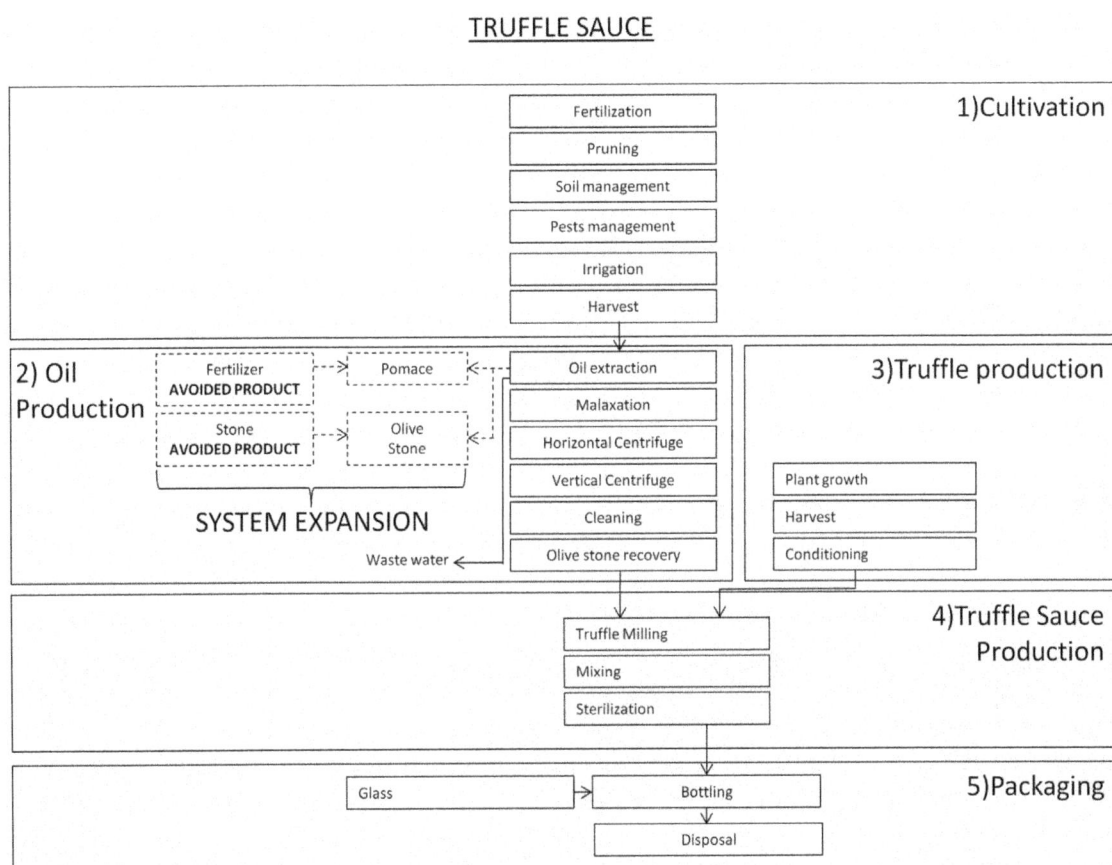

Figure 7. System boundaries [73]

Results of the analysis are proposed in Table 8. Cultivation gives an important contribution to the final impact of truffle sauce, while truffle production has a reduced impact, because it is a very extensive production. Olive trees cultivation uses fertilizers, diesel fuel for field operations, electricity for the olives harvest, herbicides and pesticides.

Life Cycle Stage	Contribution (kg CO_2eq/kg product)
1) Cultivation	0.94
2) Milling	0.28
3) Truffle production	0.09
4) Sauce production	0.06
5) Packaging production & disposal	0.77
6) Transport	0.03
7) Avoided emissionsAvoided fuel	0.18
- Avoided fertilizer	0.02
Total	1.93

Table 8. Carbon footprint of truffle sauce [73]

Author details

Francesco Fantozzi and Pietro Bartocci*

*Address all correspondence to: bartocci@crbnet.it

Department of Engineering, University of Perugia, Perugia, Italy

References

[1] Lash, J.; Wellington, F. Competitive advantage on a warming planet. Harv. Bus. Rev. 2007, 85 (3), 94–104.

[2] Hammond, G. Time to give due weight to the carbon footprint issue. Nature. 2007, 445 (7125), 256–256.

[3] Hertwich, E.; Peters, G. Carbon footprint of nations: A global, trade-linked analysis. Environ. Sci. Technol. 2009, 43, 6414–6420.

[4] ISO. ISO 14040: Environmental Management - Life Cycle Assessment - Principles and Framework. International Organization for Standardization, Geneva, Switzerland, 2006a.

[5] ISO. ISO 14044: Environmental Management - Life Cycle Assessment - Requirements and Guidelines. International Organization for Standardization, Geneva, Switzerland, 2006b.

[6] ISO. ISO 14064-1: Greenhouse Gases - Part 1: Specification with Guidance at the Organization Level for Quantification and Reporting of Greenhouse Gas Emissions and

Removals. International Organization for Standardization, Geneva, Switzerland, 2006c.

[7] ISO. ISO 14064-2: Greenhouse Gases - Part 2: Specification with Guidance at the Project Level for Quantification, Monitoring and Reporting of greenhouse Gas Emission Reductions or Removal Enhancements. International Organization for Standardization, Geneva, Switzerland, 2006d.

[8] ISO. ISO 14064-3: Greenhouse Gases - Part 3: Specification with Guidance for the Validation and Verification of Greenhouse Gas Assertions. International Organization for Standardization, Geneva, Switzerland, 2006e.

[9] ISO. ISO/TS 14067: Greenhouse Gases e Carbon Footprint of Products e Requirements and Guidelines for Quantification and Communication. International Organization for Standardization, Geneva, Switzerland, 2013.

[10] BSI. PAS 2050:2011 Specification for the Assessment of the Life Cycle Greenhouse Gas Emissions of Goods and Services. British Standards Institution, London, 2011.

[11] Technical Guidance for calculating scope 3 emissions, World Resources Institute & World Business Council for Sustainable Development, 2013.

[12] Muthu, S.S. Assessment of Carbon Footprint in Different Industrial Sectors, Volume 1, Springer, 2014, ISBN 978-981-4560-40-5.

[13] Sheehan, J.; Camobreco, V.; Duffield, J.; Graboski, M.; Shapouri, H. Life cycle inventory of biodiesel and petroleum diesel in use in an urban bus. A report prepared for US Department of Agriculture and US Department of Energy, NREL/SR-580-24089 UC Category 1503, 1998.

[14] Sims, R.E.H.; Rogner, H.H.; Gregory K. Carbon emission and mitigation cost comparisons between fossil fuel, nuclear and renewable energy resources for electricity generation. Energy Policy 2003, 31, 1315–1326.

[15] IEA. World Energy Outlook—1998 Update. International Energy Agency Report, IEA/OECD, Paris, France,1998.

[16] Glaeser, E L.; Kahn, M.E. The greenness of cities: Carbon dioxide emissions and urban development. J. Urban Econ. 2010,67, 404–418.

[17] http://www.biomassenergycentre.org.uk/portal/page?_pageid=75,163182&_dad=portal&_schema=PORTAL.

[18] Johnson, E.P. Air-source heat pump carbon footprints: HFC impacts and comparison to other heat sources. Energy Policy. 2011,39, 1369–1381.

[19] IEA).I. E. A. World Energy Outlook 2011: Are We Entering a Golden Age of Gas?; International Energy Agency: Paris, 2011

[20] EIA. In Review of Emerging Resources: US Shale Gas and Shale Oil Plays. EIA, Ed.; Washington, DC, 2011, 4.

[21] EIA. Annual Energy Outlook 2011; US Energy Information Administration: Washington, DC, 2011.

[22] Deutch, J. The natural gas revolution and its consequences. Foreign Affairs 2011, 90 (1), 82-93.

[23] Thomas, C.K. The economic impact of shale gas extraction: A review of existing studies. Ecol. Econ. 2011, 70 (7), 1243-1249.

[24] Santoro, R.; Howarth, R.W.; Ingraffea, T. Life cycle greenhouse gas emissions inventory of Marcellus shale gas. Technical Report of the Agriculture, Energy, & Environment Program, Cornell University, Ithaca, NY, 2011. To be archived and made available on-line.

[25] Hayhoe, K.; Kheshgi, H.S.; Jain, A.K.; Wuebbles, D.J. Substitution of natural gas for coal: Climatic effects of utility sector emissions. Clim. Change. 2002, 54, 107–139.

[26] Wood, R.; Gilbert, P.; Sharmina, M.; Anderson, K.; Fottitt, A.; Glynn, S.; Nicholls, F. Shale Gas: A Provisional Assessment of Climate Change and Environmental Impacts. Tyndall Center, University of Manchester, Manchester, England, 2011. http://www.tyndall.ac.uk/sites/default/files/tyndallcoop_shale_gas_report_final.pdf.

[27] Howarth, R.W.; Santoro, R.; Ingraffea, A. Methane and the greenhouse-gas footprint of natural gas from shale formations. Clim. Change, 2011, 106, 679–690. DOI 10.1007/s10584-011-0061-5.

[28] Cherubini, F.; Bird, N.D.; Cowie, A.; Jungmeier, G.; Schlamadinger, B.; Woess-Gallasch, S. Energy- and Greenhouse Gas-based LCA of Biofuel and Bioenergy Systems: Key Issues, Ranges and Recommendations. Resou. Conser. Recy. 2009, 53, 434–447.

[29] Ackermann, T.; Garner, K.; Gardiner, A. Wind power generation in weak grids—economic optimisation and power quality simulation. Proceedings of the World Renewable Energy Congress, Perth, Australia, Murdoch University, 1999. pp. 527–532, ISBN 0-86905-695-6.

[30] EWEA Wind Energy—The Facts European Wind Energy Association. Report Prepared for the Directorate-General XVII Energy, European Commission, 1999.

[31] EPRI/DOE Renewable Energy Technology Characterizations. Electric Power Research Institute and US Department of Energy, Report EPRI TR-109496,December, 1997.

[32] KPMG. Solar Energy—From Perennial Promise to Competitive Alternative. Project Number 562. KPMG Bureau voor Economische Argumentatie,. Hoofddorp, Netherlands, 1999, p. 61.

[33] Neij, L. Use of experience curves to analyze the prospects for diffusion and adoption of renewable energy technology. Energy Policy, 1997, 23, 1099–1107.

[34] AGO, Australian Greenhouse Office. Renewable Energy Showcase Projects. Australian Greenhouse Office, Canberra, 1998. www.greenhouse.gov.au/renewable/renew3.html.

[35] Lund, H.; Moller, B.; Mathiesen, B.V.; Dyrelund, A. The role of district heating in future renewable energy systems. Energy 2010, 35, 1381–1390.

[36] EC. Environmental Impact of Products (EIPRO): Analysis of the Life Cycle Environmental Impacts Related to the Total Final Consumption of the EU 25. European Commission Technical Report EUR 22284 EN, 2006.

[37] SEPA. Konsumtionens klimatpåverkan (The Climate Impact of Consumption). Report No 5903. Swedish Environmental Protection Agency, Stockholm, 2008.

[38] Garnett, T. Where are the best opportunities for reducing greenhouse gas emissions in the food system (including the food chain)? Food Pol.2011, 36, 23–32.

[39] IPCC. Contribution of Working Group I to the Fourth Assessment Report of the Intergovernmental Panel on Climate Change. Cambridge University Press, Cambridge, 2007.

[40] Tesco. Product carbon footprint summary. 2012. http://www.tescoplc.com/assets/files/cms/ Tesco_Product_Carbon_Footprints_ Summary(1).pdf. Accessed 22 May 2013.

[41] Lantmännen. Klimatdeklarationer. (Climate Declarations). 2013. http://lantmannen.se/omlantmannen/press–media/publikationer/klimatdeklarationer/. Accessed 16 May 2013.

[42] MAX. Klimatdeklaration (Climate Declaration). 2013. http://max.se/sv/Maten/Klimatdeklaration/. Accessed 22 May 2013.

[43] Roy, P.; Nei, D.; Orikasa, T. et al. A review of life cycle assessment (LCA) on some food products. J. Food Eng. 2009, 90, 1–10.

[44] de Vries, M.; de Boer, I.J.M. Comparing environmental impacts for livestock products: A review of life cycle assessments. Livestock Sci. 2010, 128, 1–11.

[45] Nijdam, D.; Rood, T.; Westhoek, H. The price of protein: review of land use and carbon footprints from life cycle assessments of animal food products and their substitutes. Food Pol. 2012, 37, 760–770.

[46] Röös, E.; Sundberg, C.; Tidåker, P.; Strid, I.; Hansson, P.-A. Can carbon footprint serve as an indicator of the environmental impact of meat production? Ecol. Ind. 2013, 24, 573–581.

[47] Röös, E. Mat-klimat-listan Version 1.0 (The Food-Climate-List Version 1.0) Report 2012:040. Department of Energy and Technology, Swedish University of Agricultural Sciences, Uppsala, 2012.

[48] Beauchemin, K.; Kreuzer, M.; O'Mara,C.; McAllister, T. Nutritional management for enteric methane abatement: A review. Aust. J. Exp. Agri. 2008, 48, 21–27.

[49] Shibata, M.; Terada, F. Factors affecting methane production and mitigation in ruminants. Anim. Sci. J. 2010, 81, 2–10.

[50] Flysjö, A. Greenhouse Gas Emissions in Milk and Dairy Product Chains—Improving the Carbon Footprint of Dairy Products. Dissertation, Aarhus University, 2012.

[51] Beddington, J.; Asaduzzaman, M.; Fernandez, A.; et al. Achieving Food Security in the Face of Climate Change: Summary for Policy Makers from the Commission on Sustainable Agriculture and Climate Change. CGIAR Research Program on Climate Change, Agriculture and Food Security (CCAFS), Copenhagen, 2011.

[52] Foley, J.; Ramankutty, N.; Brauman, K.A.; et al. Solutions for a cultivated planet. Nature 2011, 478, 337–342.

[53] Foresight. The Future of Food and Farming. Executive Summary. The Government Office for Science, London, 2011.

[54] SBA. Ett klimatvänligt jordbruk 2050 (Climate Friendly Agriculture 2050). Report 2050:35. Swedish Board of Agriculture, Jönköping, 2012.

[55] Ahlgren, S. Crop Production Without Fossil Fuel. Production Systems for Tractor Fuel and Mineral Nitrogen Based on Biomass. Dissertation, Swedish University of Agricultural Sciences, 2009.

[56] Thomassen, M.; Dalgaard, R.; Heijungs, R.; de Boer, I. Attributional and consequential LCA of milk production. Int. J. LCA 2008, 13, 339–349.

[57] Strid, E.I.; Elmquist, H.; Stern, S.; Nybrant, T. Environmental systems analysis of pig production. The impact of feed choice. Int. J. LCA 2005, 10, 143–154.

[58] Pelletier, N.; Pirog, R.; Rasmussen, R. Comparative life cycle environmental impacts of three beef production strategies in the Upper Midwestern United States. Agri. Syst. 2010, 103, 380–389.

[59] van der Velden, N.M.; Patel, M.K.; Vogtländer, J.G. LCA benchmarking study on textiles made of cotton, polyester, nylon, acryl, or elastane. Int. J. LCA 2014, 19, 331–356. DOI 10.1007/s11367-013-0626-9.

[60] Collins, M.; Aumônier, S. Streamlined Life Cycle Assessment of Two Marks & Spencer Plc Apparel Products. Environmental Resources Management, Oxford, 2002.

[61] Kalliala, E.; Talvenmaa, P. Environmental profile of textile wet processing in Finland. J. Clean. Prod. 1999, 8, 143–154.

[62] Koç, E.; Çinçik, E. Analysis of energy consumption in woven fabric production. Fibres Text East Eur. 2010, 18 (79), 14–20.

[63] Shen, L. Bio-based and Recycled Polymers for Cleaner Production— An Assessment of Plastics and Fibres. Ph.D. Thesis, Department of Science, Technology and Society (STS)/Copernicus Institute, Utrecht University, 2011.

[64] Laursen, S.E.; Hansen, J.; et al. Environmental Assessment of Textiles. Environmental Project No. 369, Danish Environmental Protection Agency, 1997.

[65] Walser, T.; Demou, E.; Lang, D.J.; Hellweg, S. Prospective environmental life cycle assessment of nanosilver T-shirts. Environ. Sci. Technol. 2011, 45 (10), 4570–4578.

[66] Ecoinvent. Database Ecoinvent Version v2.2. The Swiss Centre for Life Cycle Inventories, 2010.

[67] Humphreys, K.; Mahasenan, M. Towards a Sustainable Cement Industry – Substudy 8: Climate Change. World Business Council for Sustainable Development: Cement Sustainability Initiative, 2002.

[68] Schneider, H.Y.; Samaniego, J. La huella del carbono en la producción, distribución y consumo de bienes y servicios. Naciones Unidas, CEPAL, Santiago de Chile, 2009, 46 pp.

[69] Cagiao, J.; Gómez, B.; Doménech, J.L.; Gutiérrez Mainar, S.; Gutiérrez Lanza, H. Calculation of the corporate carbon footprint of the cement industry by the application of MC3 methodology. Ecol. Indic. 2011, 11, 1526–1540.

[70] Doménech, J.L. La huella ecológica empresarial: el caso del puerto de Gijón. In: Actas de él VII Congreso Nacional de Medio Ambiente, 22–26 Noviembre, Madrid, 2004.

[71] Doménech, J.L. Huella ecológica y desarrollo sostenible. In: AENOR Ediciones, Madrid, 2007.

[72] Carballo Penela, A.; García-Negro, M.C.; Doménech, J.E. A methodological proposal for the corporate carbon footprint: an application to a wine producer company in Galicia (Spain). Sust. J. 2009, 1, 302–318.

[73] Fantozzi, F.; Bartocci, P.; D'Alessandro, B.; Testarmata, F.; Fantozzi P. Carbon footprint of truffle sauce in central Italy by direct measurement of energy consumption of different olive harvesting techniques. J. Clean. Prod. 2015, 87, 188-196.

5

GHG Emissions from Livestock: Challenges and Ameliorative Measures to Counter Adversity

Pradeep Kumar Malik, Atul Purushottam Kolte,
Arindam Dhali, Veerasamy Sejian,
Govindasamy Thirumalaisamy, Rajan Gupta and
Raghavendra Bhatta

Additional information is available at the end of the chapter

Abstract

Livestock and climate change are interlinked through a complex mechanism and serve the role of both contributor as well as sufferer. The livestock sector is primarily accountable for the emission of methane and nitrous oxide. Methane emission takes place from both enteric fermentation and manure management; whilst nitrous oxide emission is purely from manure management. Rumen methanogenesis due to emission intensity and loss of biological energy always remains a priority for the researchers. Greenhouse gas (GHG) emissions from manure are determined by storage conditions and the organic content of the manure waste. Due to large livestock population, India is a major contributor of enteric methane emission, while its contribution to the excrement methane is negligible. In this chapter, information pertaining to enteric methane emission, excrement methane and nitrous oxide emissions and ameliorative/precautionary measures for reducing the intensity of emissions have been compiled and presented.

Keywords: greenhouse gas, GHG mitigation, livestock, methane, nitrous oxide

1. Introduction

Annual greenhouse gas (GHG) emission in 2005 was about 49 gigatonnes (Gt), wherein China contributed the maximum, followed by the United States of America and the European Union

27 [1]. The contribution of India to the total emission is about 4.25% (**Figure 1**). Worldwide livestock are integral component of agriculture and support the livelihood of billions by fulfilling 13% of energy and 28% of protein requirement. Due to the rapid change in food habits, the global demand for milk, meat and eggs in 2050 with reference to year 1990, is expected to increase 30, 60 and 80%, respectively. This additional demand will be met from livestock either by increasing their number or by intensifying productivity. The bovine and ovine population is expected to grow up at a rate of 2.6 and 2.7%, respectively, during next 35 years.

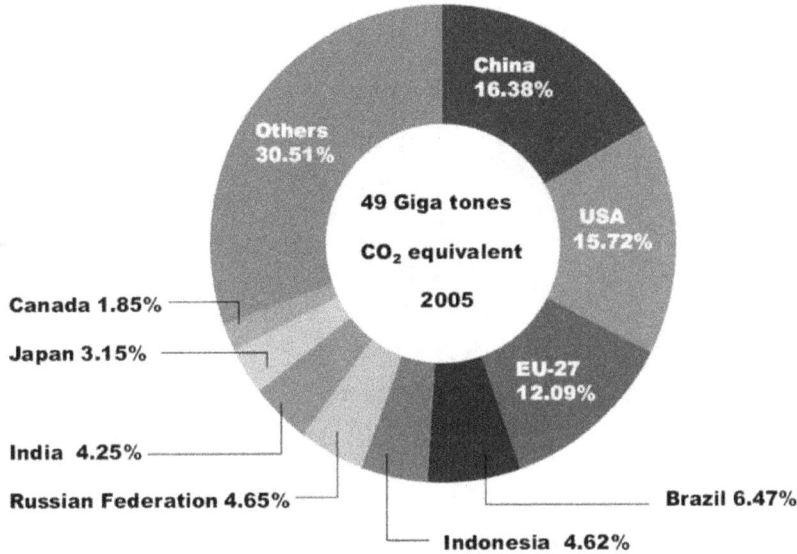

Figure 1. Nation wise greenhouse gas emissions [2] (Reprinted with permission from Takahashi [2]).

Livestock and climate change are inter-hooked in a complex mechanism where adversity of one affects another. Adverse impact of climate change on livestock across the globe will be stratified in accordance with the prevailing agro-climatic conditions. The climatic variation influences livestock in both direct and indirect ways and alterations in ambience (stresses), qualitative and quantitative changes in fodder crops, health are few of them. We can consider the livestock as one of the culprit for climate change and also the sufferer due to negative consequences of changing climate on the productive and reproductive performances of the animal. Elaborating the adverse impact of climate change on livestock production is beyond the scope of chapter and discussed elsewhere in the book. This chapter would focus primarily on the role of livestock in greenhouse gas emissions and ameliorative/precautionary measures for countering the adverse impact.

2. GHG emissions from livestock

Carbon dioxide (CO_2), methane (CH_4), and nitrous oxide (N_2O) are three major GHG emissions from livestock into the atmosphere. However, CO_2 being the part of continuous biological system cycling is not taken into consideration while calculating total GHG emission from livestock [3]. After power and land use change, agriculture including livestock is the third sector responsible for largest greenhouse gases emission. GHG emissions from different sectors are presented in **Figure 2**. Agriculture as such contributes 14% to the global GHG emissions. Of the total agricultural emissions, 38% is contributed from the soil where N_2O is one of the major GHG. GHG emission from enteric fermentation is also equally large and constitutes 32% of the total GHG emission from agriculture (**Figure 3**). In addition, rice cultivation, biomass burning, and manure management also contribute significantly and make about 30% of the agricultural emissions.

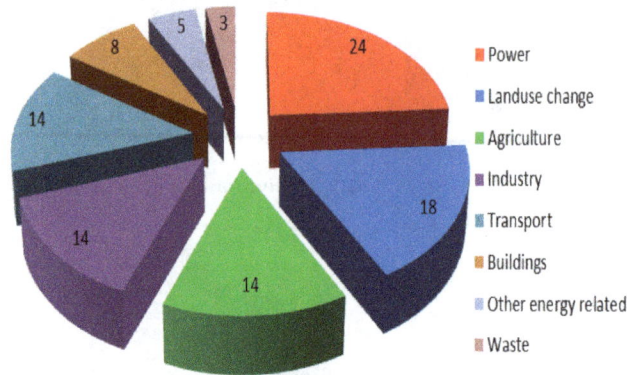

Figure 2. Sector wise GHG emissions.

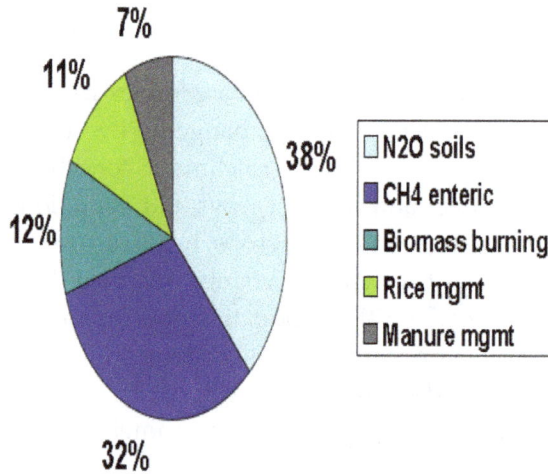

Figure 3. Global agricultural GHG emission.

Livestock emits methane both from enteric fermentation and from manure management; whilst nitrous oxide emission is purely associated with the manure management system. However, methane emission from manure management is far less than the emission from enteric fermentation. Methane emission from excrement is mainly confined to animal management operations where excrement is handled in liquid based systems. N_2O emission from manure management varies significantly between types of management system and also related to indirect emissions from other forms of nitrogen. Of the total anthropogenic methane and nitrous oxide emissions, livestock globally contribute 35 and 65% of the respective GHGs. Latin America occupies first position (23%) in the list of top enteric methane emitting countries (**Figure 4**), while Africa (14%) and China (13%) hold second and third positions. India stands at the fourth position and is accountable for 11% of the worldwide enteric methane emission (**Figure 4**). The contribution from Middle East and Eastern Europe is negligible and contributes only 2.8% of the total emission [4]. The United States' Environmental Protection Agency [5] projected that the enteric methane emission will substantially increase in 2020 and 2030 in comparison to 2010 (**Figure 5A**). Similarly, projections also imply an increase in enteric methane emission from Indian livestock than that was in 2010. However, methane and nitrous oxide emission will almost remain stabilized for the next 10–20 years (**Figure 5**).

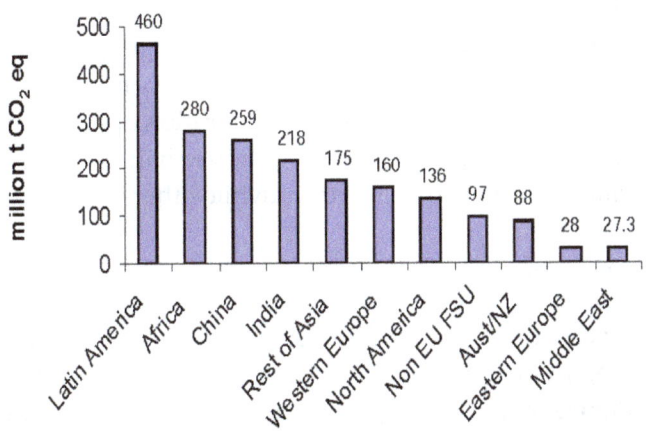

Figure 4. Region wise enteric methane emission [4].

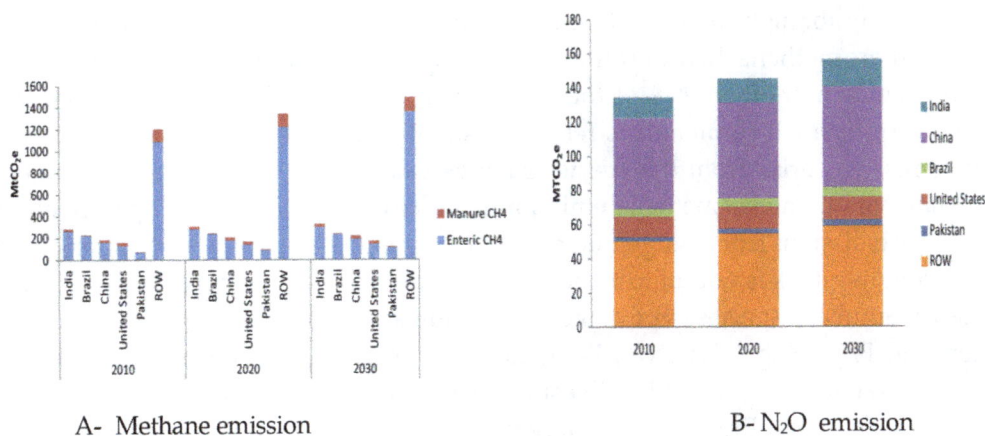

A- Methane emission   B- N_2O emission

Figure 5. Projections for 2020 and 2030 [5]. (A) Methane emission. (B) N_2O emission.

2.1. Rumen methanogenesis: good and bad associated with it

Rumen harbours a diverse group of microbes that undertake different functions from complex carbohydrate degradation to the removal of end metabolites arise from fermentation. These microbes work in a syntrophic fashion under strict anaerobic conditions and help each other in performing their functions. H_2 is a central metabolite produced in large volume from fermentation and need to be disposed off away from the rumen. Many hydrogenotrophic pathways, such as methanogenesis, reductive acetogenesis, sulfate reduction, and nitrate reduction, have been described as a sink for H_2 in the rumen. Under normal rumen functioning, methanogenesis due to the thermodynamic efficiency is the most prominent hydrogenotrophic pathway. In methanogenesis, H_2 is used for the reduction of CO_2 and conversion into methane which later on eructate from the rumen. Methanogenesis removes unwanted and fatal products of fermentation from the rumen, therefore, it is an essential pathway for the normal rumen functioning, involving the residing microbes and the host animal. The methane energy value is 55.65 MJ/kg [6] and therefore its removal deprives the host animal from a substantial fraction of ingested biological energy. This loss generally lies in the range of 6–12% of the intake [7]. In addition, enteric methane emission due to its high global warming potential (25 times of CO_2) also contributes significantly to the global warming [5]. Due to many intact disadvantages with enteric methane emission, its amelioration up to a desirable extent is much more important than any other GHG. Its relatively shorter half-life offers added opportunity to stabilize global warming in short time and meanwhile other GHG could also be tackled.

2.2. Enteric methane emission: Indian scenario

Various agencies reported quite variable figures for enteric methane emission from Indian livestock. Many have reported annual emission as high as 18 Tg per year, while others have estimated only 7 Tg (**Figure 6**). The average of these estimates comes around 8–10 Tg per year which constitutes about 11% of the global enteric methane emission. India possesses 512 million livestock [8] wherein cattle and buffaloes are the prominent species and make up to 60% of the total livestock in the country.

One of the reasons for high enteric methane emission from India is the larger bovine population which emits more methane than any other livestock species. On an average, cattle and buffaloes aggregately emits more than 90% of the total enteric methane emission of the country. The contribution from small ruminants is relatively small and constitutes only 7.7%. Rest of the methane emissions arise from the species such as yak and mithun, which are scattered to specific states only. Enteric methane emission from crossbred cattle is comparatively much more than the emissions from indigenous cattle (46 versus 25 kg/animal/year). Enteric methane emission from livestock is not uniform across the states and varies considerably according to the livestock numbers, species, type of feed and fodders, etc. The National Institute of Animal Nutrition and Physiology (NIANP), Bangalore has developed an inventory for state wise enteric methane emission from Indian livestock using 19th livestock census report. The NIANP estimates revealed Uttar Pradesh as the largest enteric methane emitting state of the country [9]. Other major methane emitting states in the country are Rajasthan, Madhya Pradesh, Bihar, West Bengal, Maharashtra, Karnataka and Andhra Pradesh (**Figure 7**). These states altogether holds 66% of the livestock population and accountable for 68% enteric methane emissions. Due to large contribution, these states can be considered as hotspots for reducing enteric methane emissions from livestock and are given priority for tackling the emission.

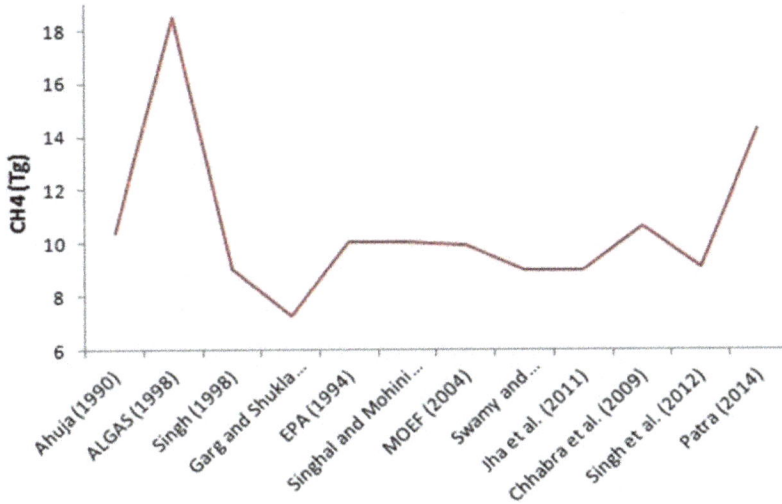

Figure 6. Disparity in enteric methane emission from Indian livestock.

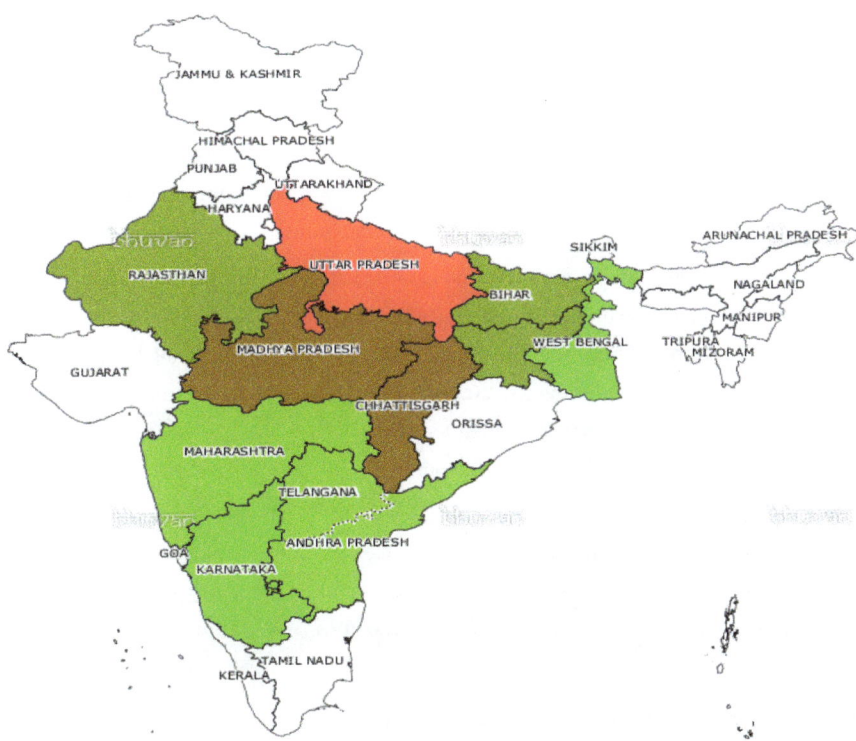

Figure 7. Major enteric methane emitting states in India.

2.3. Enteric methane amelioration: challenges and opportunities

Attempting enteric methane mitigation without understanding necessity, knowing exact emission from country/state, extent and feasibility of reduction, complexity of ruminal microbes and their syntrophic relationship will not serve the effective and sustainable reduction in long term as learnt from past experience in many countries. Archaea in the rumen are methane producing microbes. Earlier methanogens were considered under bacterial domain (prokaryotes), but recent classification by Woese [10] placed them in a distinct domain, which is remarkably different from bacteria. Methanogens archaea are primarily hydrogeno-trophic microbes, which utilize H_2 as the main substrate for methanogenesis. Though, they can Use other substrates also for methanogenesis, but H_2 remains a central metabolite and its partial pressure determines the degree of methanogenesis [11]. Due to its main role in maintaining the redox-potential (reducing environment) of rumen, H_2 is referred as *currency of fermentation* [11]. Therefore, deep understanding of rumen archaea, their substrate require-ment and role in methanogenesis is pre-requisite for achieving sustainable reduction in methane emission. The latest metagenomic approaches served as potential tool and helped in exploring many more cultured and uncultured rumen methanogens for better understanding. The effectiveness and persistency of the ameliorative approach depends on the extent of methanogens being targeted by the approach under investigation. In spite of initial reduction, enteric methane emission usually gets back to the normal level, which is due to partial targeting of methanogen community in rumen. All possible ameliorative measures for enteric methane mitigation are presented in **Table 1**.

Measures	Opportunities/Limitation	Remarks
Reducing the livestock numbers	Due to high number of low producing or non-producing ruminants methane emission per kg of livestock product is high. Killing of such livestock is not possible due to the ban on cow slaughter in the country.	Low productive animals should be graded up with rigorous selection for improving their productivity and less enteric methane emission.
Feeding of quality fodders, concentrate	Feed interventions are the best option for methane amelioration. The uninterrupted availability is a question mark. Area under pasture and permanent fodder production declining or stagnant since last three decades. Livestock are getting their fodders from 7–8% of the arable area in the country.	Improving quality fodders availability seems unrealistic under ever increasing human population and food-feed-fuel competition scenario.
Ionophore	Selective inhibition of microbes and failure to achieve the reduction in long term are big issues. Animals turn back to normal level of emission after short time. Their use is banned in many European countries.	May be tried in rotation as well in combination for sustaining the reduction in long term.
Ration balancing	Ration balancing with feed resources available at farmer's doorstep will improve the productivity with concurrent methane reduction at a low input level.	Farmers need to be made aware about the importance of ration balancing and monetary advantages from the same.

Measures	Opportunities/Limitation	Remarks
Removal of protozoa	Removal of ciliate protozoa from the rumen results in lower methane production. May witness less fibre digestibility. It is practically impossible to maintain protozoa free ruminants.	In spite of complete removal, partial defaunation may be achieved for enteric methane reduction without affecting the fibre digestion.
Reductive acetogenesis	Thermodynamics favour methanogenesis in the rumen. The affinity of acetogens for H_2 substrate is considerably lower than methanogens. It cannot work until and unless target methanogens are absent in the rumen.	Reductive acetogenesis may be promoted by simultaneously targeting rumen archaea. This will ensure less methane with additional acetate availability for the host animal.
Use of plant secondary metabolites	Under the quality fodders deficit scenario, use of PSM as methane mitigating agents is a good option. Dose optimization and validation of methane migration potential *in vivo* on a large scale is mandatory before recommendation.	Inclusion at a safe level without affecting the feed fermentability may be a viable option for enteric methane amelioration. Studies are warranted for assessing the combined action of PSM on *in vivo* methane emission.
Nitrate/Sulfate	Nitrate and sulfate hold the potential to reduce methane emission to a greater extent. These reductive processes are thermodynamically more favourable than methanogenesis. The end product from this productive process will not have any energetic gain for the animal. Intermediate products are toxic to the host animal.	Probably slow releasing sources for these compounds will reduce the toxicity chances caused by intermediate metabolites. A safe level of inclusion must be decided and tested on large number of animals by considering all the species accountable for methane emission.
Active immunization	This approach hold the potential for substantial methane reduction provided methanogen archaea of rumen is explored to a maximum extent for identifying the target candidate for the inclusion in vaccine.	Information on the species and bio-geographic variation in methanogenic archaeal community should be explored for considering this approach for enteric methane amelioration.
Disabling of surface proteins	It is well established that methanogens adhere to the surface of other microbes for H_2 transfer through surface proteins. Identifying and disabling of these surface proteins will certainly reduce enteric methane emission by cutting the supply of H_2.	This is an unexplored area and need some basic and advance research for exploring the possibility.
Biohydrogentation	Restricting the H_2 supply to methanogens through alternate use in bio-hydrogenation, decrease enteric methane amelioration. Use of fat/lipids at a high level depresses fibre digestion. Of the total, only about 5–7% of H_2 is utilized in this process.	This approach is not practical due to high cost of fat/lipids and fibre depression at a high level of use.

Table 1. Ameliorative measures for enteric methane mitigation.

2.4. Plant secondary metabolites as ameliorating agent

Plant secondary metabolites (PSMs) are organic compounds that are not directly involved in the growth, development, or reproduction, but play an important role in plant defence against herbivores. Plant secondary metabolites, on the basis of their biosynthetic origins can be grouped into three: flavonoids, and allied phenolic and polyphenolic compounds; terpenoids and nitrogen-containing alkaloids; and sulphur-containing compounds. Among these, tannins are most important for enteric methane amelioration. Chemically, they are polyphenolic compounds with varying molecular weights, and have the ability to bind natural polymers, such as proteins and carbohydrates. Based on their molecular structure, tannins are classified as either hydrolysable tannins (HT; polyesters of gallic acid and various individual sugars) or condensed tannins (CT; polymers of flavonoids), although there are also tannins that represent combinations of these two basic structures. As PSMs are integral components of abundant phyto-sources and are required in very limited quantity for exerting anti-methanogenic action, therefore, using them as an ameliorating agent would cost very little to the stakeholders.

The tannins exert their anti-methanogenic activity through direct inhibition of methanogen archaea or indirectly by interfering with protozoa and restricting the interspecies H_2 transfer [12, 13]. More than 100 phyto-sources have been evaluated in our laboratory (*in vitro*) for determining their methane mitigation potential and to optimize their level of inclusion in the animal diet [14, 15].

Saponin is another group of plant secondary metabolites that possess a carbohydrate moiety attached to an aglycone, usually steroid or triterpenoid. Saponins are widely distributed in the plant kingdom and research revealed the use of saponin as such or as phyto source legumes that contain an appreciable amount of saponins. Malik and Singhal [16] in an *in vitro* study reported 29% reduction in methane production on the addition of 4% commercial grade saponin in wheat straw and concentrate based diet. Further, same authors [17] also reported a reduction of 21% in enteric methane emission in Murrah buffalo calves due to the supplementation of saponin-containing lucerne fodder as 30% of the diet. In an *in vitro* study, Malik et al. [18] observed a significant reduction in methane production due to the supplementation of first cut alfalfa fodder. The addition of saponin or saponin-containing fodder affects methanogenesis primarily through the anti-protozoa action or altering the fermentation pattern and direct inhibition of rumen methanogens [19].

3. GHG emissions from manure management

Livestock manure proved a valuable material that contains required nutrients for plant growth and an excellent soil amendment for improving soil quality and health. Methane is a major greenhouse gas emitted from manure during anaerobic decomposition of the organic matter. Another important greenhouse gas is nitrous oxide, which contrarily emits from aerobic storage of excrement. A pictorial presentation of the possible sources for methane and nitrous oxide emission is provided in **Figure 8**. The thick arrow in **Figure 8** represents the major source for a particular GHG.

Figure 8. Sources of GHG from livestock excrement.

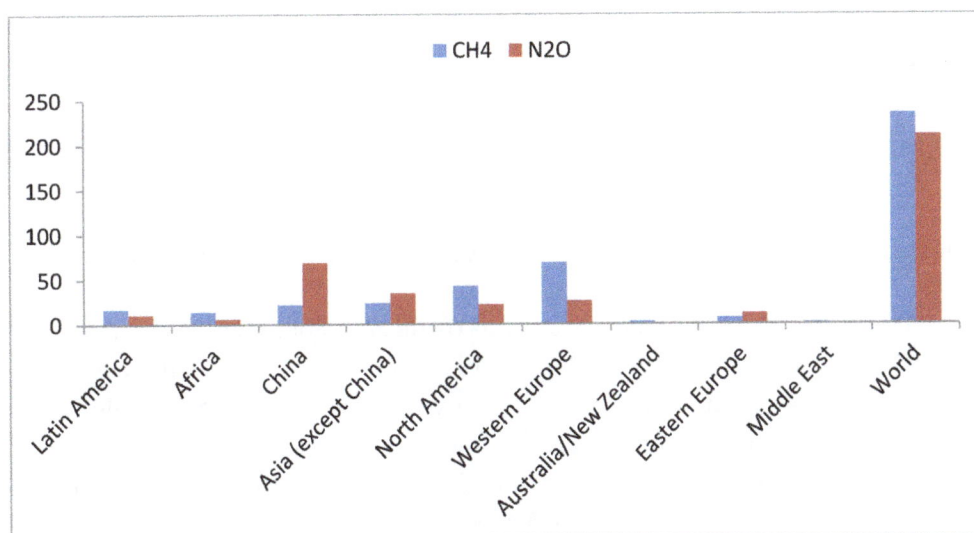

Figure 9. Methane and nitrous oxide emission from manure management in different regions of the world [22] (modified with permission from EPA [3]; O'Mara [21]; UNEP [22]).

The extent of emission of particular greenhouse is determined by the disposal and processing of waste. For example, methane is the primary GHG emit from the excrement, if waste is flushed with water and stored in lagoon; while on the other hand, nitrous oxide is the primary

GHG, if waste is stored as heap in an aerobic environment (**Figure 8**). Methane emission from livestock excrement as such is not a major issue in developing countries, like India. However, excrement is a major source of methane emission in developed world, where excrement is mainly disposed anaerobically. Worldwide production of methane and nitrous oxide annually contribute about 235 and 211 Mt of $CO_2.$eq, respectively [20, 21]. Regional estimates of manure methane and nitrous oxide are presented in **Figure 9**. Asian countries due to following aerobic storage of excrement contribute about 49% of the total nitrous oxide emissions (**Figure 9**). The aerobic conditions favour nitrous oxide emission from excrement and disfavour methanogenesis. The contribution from America and Africa to total nitrous oxide emission is 15 and 3%, respectively. On the other hand, methane emission from manure is highest in America (22%), which is obviously due to anaerobic processing of animal wastes.

Methane	Manure Methane (kg x 10^5)		
	Estimated	Projected	
	2010	2025	2050
World	11,414	12,849	15,046
India	1096	1221	1543
% of total	9.6	9.5	10.2
Methane	Manure N_2O (kg x 10^5)		
	Estimated	Projected	
	2010	2025	2050
World	383	445	516
India	15.3	17.5	21.4
% of total	3.9	3.9	4.1

Table 2. Estimate and projected emissions of methane and methane from manure management [23].

Patra [23] has estimated the methane and nitrous oxide emissions from manure management and also made projections for 2025 and 2050 (**Table 2**). He projected a small increase from 9.6 to 10.2% to the manure methane emission in India over a period of 30 years (**Table 2**). Likewise a small increase is also projected for manure nitrous oxide emission from both world and India. He projected an increase of 133 Mt $CO_2.$eq nitrous oxide from total manure produced in the world; while in India it would be around 6 Mt $CO_2.$eq between 2010 and 2030.

The type and quantity of diet are deciding factors for the extent of methane emission from a given volume of manure [24]. International Panel on Climate Change (IPCC) proposed a value of 0.24 L methane per gram of volatile solids (VSs) for dairy cattle [25]. Hashimoto et al. [26] evaluated the methane emission from manure of beef cattle fed different quantities of corn silage and corn grain in the following percentage: 92–0%, 40–53% and 7–88%, respectively. The corresponding emission figures were 0.173, 0.232 and 0.290 L per gram of VS, respectively.

Manure management is an essentiality to be considered for minimizing GHG emissions from excrement processing. The decomposition of dung under anaerobic conditions produces methane. Anaerobic conditions usually arise when dung is mainly disposed along with liquid. Total dung produced and the fraction that undergoes anaerobic decomposition influence methane emissions. When manure is stored or treated as a liquid in lagoons, ponds, tanks or pits, it decomposes anaerobically and produces significant methane. The temperature and the retention in storage vat greatly affect the degree of methanogenesis. Handling dung in the solid form (e.g. stacks or heap) or deposition in pasture and rangelands, accelerate the aerobic decomposition and hence, produce very less methane. The methane production from dung depends on its VS content. VS are organic content of dung which contains both biodegradable and non-biodegradable fractions. VS excretion rates may be retrieved from the literature or determined by conducting experiments. Enhanced characterisation methods can be used for estimating the VS content [Equation 1] . The VS content of dung is considered equivalent to the undigested fraction of the diet, which is consumed but not digested and therefore, excreted as faeces. VS excretion rate may be worked out using the equation of Dong et al. [27]

Volatile solid excretion rates [27],

$$VS = \left[GE\left(1 - \frac{DE\%}{100}\right) + (UE\ GE) \right]\left[\left(\frac{1-ASH}{18.45}\right) \right]. \tag{1}$$

Using the VS excretion rate, the methane emission factor from dung may be determined as per The equation 2 given below [27]:

$$EF_{(T)} = \left(VS_{(T)}\ 365\right)\left[B_{o(T)}\ 0.67\,kg\,/\,m^3 \sum_{S,k} \frac{MCF_{S,k}}{100}\ MS_{(T,S,k)} \right]. \tag{2}$$

Nitrous oxide emissions from manure management directly arise from the nitrification and denitrification process. The extent of nitrous oxide emission from manure during storage depends on nitrogen and carbon contents as well as storage duration. Nitrification, that is, oxidation of ammonia nitrogen to nitrate nitrogen, is a necessary step in the generation of nitrous oxide from animal manures. Nitrification occurs when stored dung has sufficient supply of oxygen. During denitrification, which is an anaerobic process, nitrites and nitrates are converted into nitrous oxide and dinitrogen. Direct nitrous oxide emission from manure management may be estimated using following equation:

Direct nitrous oxide emission from manure management [27]:

$$N_2O_{D(mm)} = \left[\sum_S \left[\sum_S \left(N_{(T)}\ Nex_{(T)}\ MS_{(T,S)} \right) \right] EF_{3(S)} \right] \frac{44}{28}. \tag{3}$$

3.1. Measures for reducing GHG

Precautionary or ameliorative measures to ensure less greenhouse gas emission from manure depend on the storage conditions. Due to contradictory environmental conditions required for methane and nitrous oxide emissions, similar mitigating or precautionary measures cannot tackle both the gases at the same time. Therefore, we should fix the priority before attempting the mitigation and process the excrement accordingly. For mitigating methane and nitrous oxide emissions from manure management, few precautionary/ameliorative measures are furnished in **Table 3**.

GHG	Measures
Methane	• Handling of manure in the solid form or deposition on pasture rather than storing it in a liquid based system. However, this may increase nitrous oxide emission.
	• Capturing methane from manure decomposition for producing renewable energy.
	• Avoid adding straw to manure which serve as a substrate for anaerobic bacteria.
	• Application of manure to soil as early as possible to avoid the anaerobic storage of manure which encourages anaerobic decomposition and favour methanogenesis.
	• Application of manure when soil surface is wet should be avoided as it may lead to increase methane emissions.
	• Improve animal's feed conversion efficiency either by feeding quality feeds or by processing to decrease GHG emissions.
	• Cover lagoons with plastic covers or any other means to capture GHGs.
N_2O	• Manure should apply shortly before crop growth for efficient utilization of available nitrogen by crop.
	• Avoid applying manure in winter as it can lead to high emission.
	• Hot and windy weather should be avoided for applying manure because these conditions can increase nitrous oxide emissions.
	• Follow the ideal practices for improving drainage, avoiding soil compaction, increasing soil aeration, and use nitrification inhibitors.
	• Even application of manure around the pasture.
	• Maintain healthy pastures by implementing beneficial management grazing practices to help increase the quality of forages.
	• Include low protein levels and the proper balance of amino acids in the diet to minimize the amount of nitrogen excreted, particularly in urine. Use phase feeding to match diet to growth and development.
	• Storage underground surface with lower temperatures reduces microbial activities.

Table 3. Precautionary/ameliorative measures for reducing GHG emissions from manure management.

4. Summary

Livestock are the major source for anthropogenic GHG emissions as they tend to emit methane from enteric fermentation and manure management and nitrous oxide from manure management. These GHGs as compared to carbon dioxide have very high global warming potential. Apart from accelerating the global warming, enteric methane emission from livestock also carry off substantial fraction of the energy which is supposed to be used by the host animal. A country like India cannot afford this energy loss, as it demands additional feed resources to compensate the loss. The adoption of mitigation options for enteric methane amelioration should be based on the feasibility of intervention(s) in a specific region. Our focus should be on those approaches which may persist in a long run and lead to 20–25% reduction in enteric methane emission. Methane and nitrous oxide emissions from manure management demands different storage conditions. Due to storage conditions (mainly aerobic), the methane emission from manure in the developing countries is not very alarming and hence, our focus should be on reducing nitrous oxide emission from manure management by developing the interventions which at least ensure that nitrous oxide emission has not gone up while trying to mitigate methane emission from manure management.

Author details

Pradeep Kumar Malik[1*], Atul Purushottam Kolte[1], Arindam Dhali[1], Veerasamy Sejian[1], Govindasamy Thirumalaisamy[1], Rajan Gupta[2] and Raghavendra Bhatta[1]

*Address all correspondence to: malikndri@gmail.com

1 ICAR-National Institute of Animal Nutrition and Physiology, Bangalore, India

2 Indian Council of Agricultural Research, New Delhi, India

References

[1] WRI: Climate Analysis Indicators Tool (CAIT), version 9.0. 2011; World Resource Institute, Washington DC, USA

[2] Takahashi J: Perspective on livestock generated GHGs and climate. In: Malik PK, Bhatta R, Takahashi J, Kohn RA and Prasad CS (eds). Livestock production and climate change. CABI book published by CAB International UK and USA; 2015. pp. 111–124.

[3] EPA, Holtkamp J, Hayano D, Irvine A, John G, Munds Dry O, Newland T, Snodgrass S, Williams M: Inventory of U.S. greenhouse gases and sinks: 1996–2006. Environmental

Protection Agency, Washington, DC; 2006, http://www.epa.gov/climatechange/emissions/ downloads/08_Annex_1-7.pdf

[4] EPA: Global mitigation of non-CO2 greenhouse gases:2010-2013. United States Environmental Protection Agency 2013. Office of Atmospheric Programs (6207J) EPA-430-R-13-011 Washington, DC.

[5] EPA: Global mitigation of non-CO2 greenhouse gases: 2010–2013. United States Environmental Protection Agency, Washington; 2013, EPA-430-R-13-011.

[6] Crutzen PJ, Aselmann I, Seiler W: Methane production by domestic animals, wild ruminants, other herbivorous fauna, and humans. Tellus. 1986; 38B: 271–284.

[7] Van Nevel CJ, Demeyer DI: Control of rumen methanogenesis. Environmental Monitoring and Assessment. 1996; 42: 73–97. DOI: 10.1007/BF00394043.

[8] 19th Livestock Census: All India Report 2012. Ministry of Agriculture, Department of Animal Husbandry, Dairying and Fisheries, Krishi Bhavan, New Delhi. p. 130.

[9] Bhatta R, Malik PK, Kolte AP, Gupta R: Annual progress report of outreach project on methane. NIANP, Bangalore, India; 2016.

[10] Woese CR, Kandler O, Wheelis ML: Toward a natural system of organisms: proposal for the domains Archaea, Bacteria, and Eucarya. Proceedings of the National Academy of Sciences USA. 1990; 87, 4576–4579.

[11] Hegarty RS, Gerdes R: Hydrogen production and transfer in the rumen. Recent Advances in Animal Nutrition. 1998; 12, 37–44.

[12] Bhatta R, Uyeno Y, Tajima K, Takenaka A, Yabumoto Y, Nonaka I, Enishi O, Kurihara M: Difference in the nature of tannins on *in vitro* ruminal methane and volatile fatty acid production and on methanogenic archaea and protozoal populations. Journal of Dairy Science. 2009; 92: 5512–5522.

[13] Hristov AN, Joonpyo Oh, Lee C, Meinen R, Montes F, Ott T, et al: Mitigation of greenhouse gas emissions in livestock production: a review of technical options for non-CO_2 emissions. In: Gerber P, Henderson B and Makkar H (eds.), FAO Animal Production and Health Paper No. 177, FAO, Rome, Italy; 2013.

[14] Bhatta R, Baruah L, Saravanan M, Suresh KP, Sampath KT: Effect of medicinal and aromatic plants on rumen fermentation, protozoal population and methanogenesis in vitro. Journal of Animal Physiology and Animal Nutrition. 2013; 97: 446-456.

[15] Bhatta R, Saravanan M, Baruah L, Sampath KT, Prasad CS: *in vitro* fermentation profile and methane reduction in ruminal cultures containing secondary plant compounds. Journal of Applied Microbiology. 2013; 115: 455–465.

[16] Malik PK, Singhal KK: Influence of supplementation of wheat straw based total mixed ration with saponins on total gas and methane production in vitro. Indian Journal of Animal Sciences. 2008; 78: 987–990.

[17] Malik PK, Singhal KK: Effect of alfalfa fodder supplementation on enteric methane emission measured by sulfur hexafluoride technique in murrah buffaloes. Buffalo Bulletin. 2016; 35: 125–134.

[18] Malik PK, Singhal KK, Deshpande SB: Effect of lucerne fodder (first cut) supplementation on *in vitro* methane production, fermentation pattern and protozoal counts. Indian Journal of Animal Sciences. 2010; 80: 998–1002.

[19] Malik PK, Bhatta R, Soren NM, Sejian V, Mech A, Prasad KS, Prasad CS: Feed-based approaches in enteric methane amelioration. In: Malik PK, Bhatta R, Takhashi J, Kohn RA and Prasad CS (eds), Livestock production and climate change. CABI Publishers, Oxfordshire, UK; 2015. pp. 336–359.

[20] EPA: Global anthropogenic non-CO2 greenhouse gases emissions: 1990–2020. United States Environmental Protection Agency, , Washington; June 2006, EPA 430-R-06-003

[21] O'Mara FP: The significance of livestock as a contributor to global greenhouse gas emission today and in the near future. Animal Feed Science and Technology. 2011; 166–167: 7–15.

[22] UNEP: Growing greenhouse gas emissions due to meat production. UNEP Global Environmental Alert Service (GEAS); 2012. Available from: http://www.unep.org/pdf/unep-geas_oct_2012.pdf

[23] Patra AK: Trends and projected estimates of GHG emissions from Indian livestock in comparison with GHG emissions from world and developing countries. Asian-Australian Journal of Animal Sciences. 2014; 27: 592–599.

[24] Masse DI, Masse L, Claveau S, Benchaar C, Thomas O: Methane emissions from manure storages. Transactions of the ASABE. 2008; 51: 1775–1781.

[25] IPCC, Watson RT, Zinyowera MC, Moss RH (eds): The regional impacts of climate change: an assessment of vulnerability. Cambridge University Press, Cambridge, UK; 1997. 517 p.

[26] Hashimoto AG, Varel VH, Chen YR: Ultimate methane yield from beef cattle manure: effect of temperature, ration constituents, antibiotics, and manure age. Agricultural Wastes. 1981: 3(4): 241–256.

[27] Dong H, Mangino J, McAllister TA, Hatfield JL, Johnson DE, Lassey KR, de Lima MA, Romanovskya A: Emission from livestock and manure management (Chapter 10). In: 2006 IPCC guidelines for National Greenhouse Gas Inventories, Volume 4: Agriculture, Forestry and Other Land Use; 2006. Available online at http://www.ipcc-nggip.iges.or.jp/public/2006gl/pdf/4_Volume4/V4_10_Ch10_Livestock.pdf

6

About the Concept of the Environment Recycling—Energy (ERE) in the Romanian Steel Industry

Adrian Ioana and Augustin Semenescu

Additional information is available at the end of the chapter

Abstract

This paper takes as its starting point an analysis of the ecological functioning of the electric arc furnace (EAF). Thus, we present a classification of emissions generated by EAF, including limits of variation in chemical composition of "dust" issued by EAF in various countries and limit values for permissible concentrations of these emissions. The paper presents and analyzes various abstraction and treatment-related emissions for hipo-polluting operation of EAF. In this chapter, the correlations between macro system represented by metallurgical environment and interacting systems: System-Energy-Recycling Environment (ERE), Ecological system (ECO), and Recycling, Reclamation System (REC-REV) are presented. These correlations are presented in the spirit of sustainable development concepts (DC) and total quality (TQ).

Keywords: environment-recycling-energy, metallurgical process technology, ecological system

1. Introduction

Reducing the amount of emissions and greenhouse gas immissions is an important environmental goal, including specific achievement to ensure the optimal concept of sustainable development.

The Electric Arc Furnace (EAF) for steel development is a powerful polluter. From this point of view, studying and optimizing the functioning of this complex metallurgical aggregate, including the conception of ERE, are of special importance. These activities of study and optimization ensure the optimal conditions for sustainable development.

Metallurgical process is a macroenvironment characterized mainly by the following systems:

- Metallurgical Process Technology system (MPT)—it is analyzed and defined by technological parameters and technological procedures applied.

- The Environment Energy Recycling System (EER)—it defines and characterizes the energy resources necessary for the transformations sources and the metallurgical processes.

- The Ecological System (ECO)—it refers to the organic processing systems pollutant outputs from MPT and EER.

- The Recycling and Revaluation System (REC-REV)—it refers to the energy and material transformations that occur within process flows. This system consists of two subsystems in turn, namely:

- The recycling subsystem (the capitalization) energy (R_E); this subsystem studies and makes more efficient the recycling of energy both within the same ecosystem and within the energetic exchanges between the ecosystems that are interdependent.

- The recycling subsystem (the capitalization) of materials (R_M); this subsystem is directly correlated with energy recycling subsystem, and it represents at the same time a qualitative and quantitative measure of it.

An ecosystem performs three important functions. These functions are as follows:

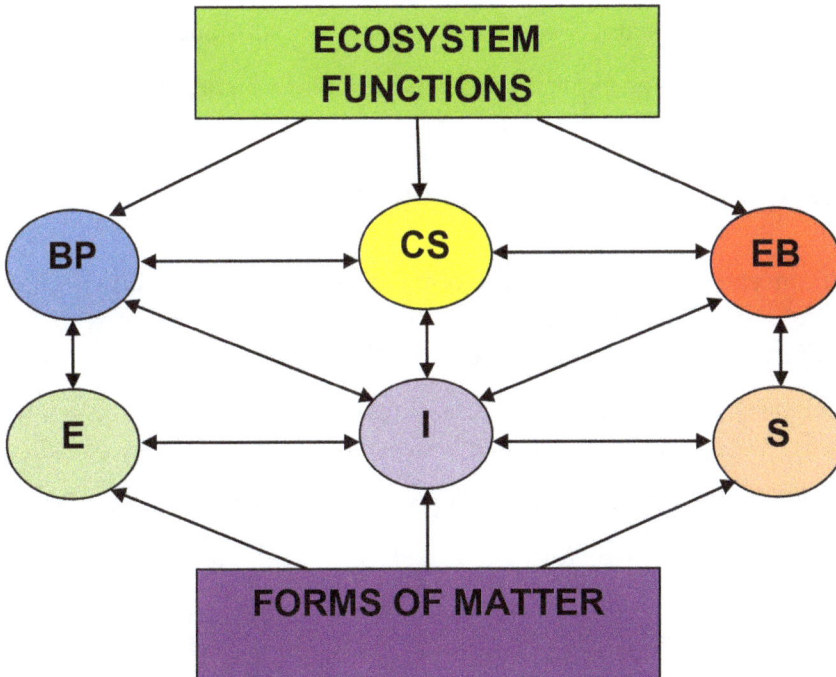

Figure 1. The correlations between an ecosystem functions and forms of matter. Source: own research. BP—the biological productivity; CS—the circulation of substance; EB—the ecological balance; E—energy; I—information; and S—substance.

- The biological productivity (BP)—the biological productivity of an ecosystem is directly dependent both on the quality of the biotope (geographical environment) and on the biocenosis (all living organisms).

- The circulation of substance (CS)—this function of the ecosystem refers to the exchange of materials both within the same ecosystem and between different ecosystems that are interdependent.

- The ecological balance (EB)—is a function of quantification (quantitative determination) of the exchanges of matter and energy within an ecosystem.

The functions of an ecosystem must be analyzed in relation to the transformations through the three forms of matter:

- Energy (E)—energy as a form of matter is a measure of its quality.

- Information (I)—the information is the genetic carrier of the matter; it also characterizes the history and evolution of the respective matter.

- Substance (S)—it is a way of quantifying the matter. **Figure 1** shows the correlations between an ecosystem functions and forms of matter.

A brief explanation is useful to the ecosystem functions:

- The transformation of energy is illustrated by the size and variation in biological productivity.

- The circuit of macroelements reflects the changes of the substance.

- The ecological balance is an expression of the size of information.

2. The correlations of MPT-ERE-ECO-REC-REV

Based on the concepts of sustainable development (SD) and total quality (TQ), the effective analysis of a process and metallurgy must put on the forefront quantify correlations MPT-EER-ECO-REC-REV. In **Figure 2**, we present these correlations scheme.

The ecosystem, by definition, is a group consisting of biotopes, which sets a whole different relationships, both between organisms and between organisms and between abiotic factors.

This first definition of ecosystem requires a specific definition and an explanation of other concepts, such as:

- Biotope

- Biocenosis

- Abiotic

The Biotope is defined as the geographical environment in which there lives a group of living organisms (humans, plants, animals, etc.) in homogeneous conditions.

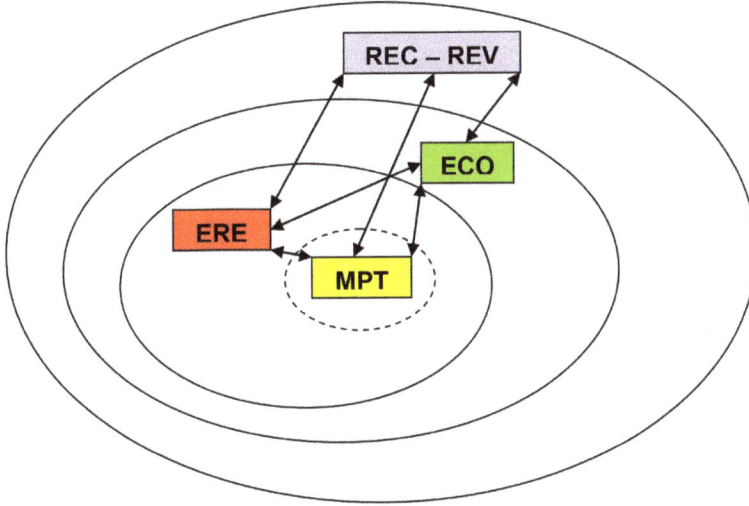

Figure 2. The correlations between the systems of MPT, EER, ECO and REC-REV to ensure concepts of DD and CT. Source: own research. MPT—The Metallurgical Process Technology; ERE—The Environment-Recycling Energy System; ECO—The Ecological system; REC–REV—The Recycling and Recovery of the System Recovery; SD—Sustainable Development; TQ—Total Quality.

The Biocenose represents all living organisms that inhabit a particular geographical environment (biotope).

The Abiotic refers to something lifeless, incompatible with life.

It is based on this first definition in **Figure 3**. This is the diagram of the ecosystem.

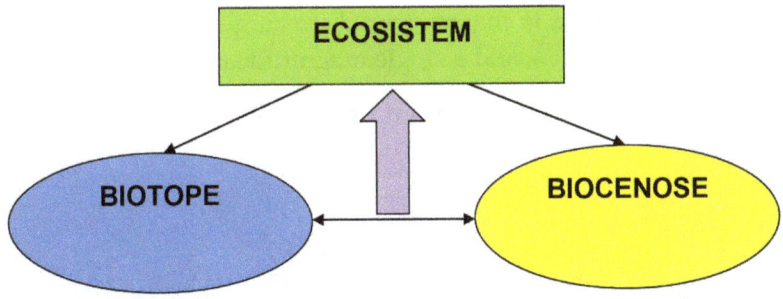

Figure 3. The diagram of the ecosystem. Source: [5].

The following general concepts on which they define in the automatization domeny is the System Ecometalurgic (SE).

The System Ecometallurgic is an ecosystem of custom-specific conditions and technologies in metallurgy (industry metallic materials—ferrous and nonferrous), characterized by a geographic environment and an specific industry (biotope) and by groups of living organisms (people, plants, and animals) that inhabit this environment (biocenosis).

Figure 4 shows a schematic diagram of the ecometalurgic system.

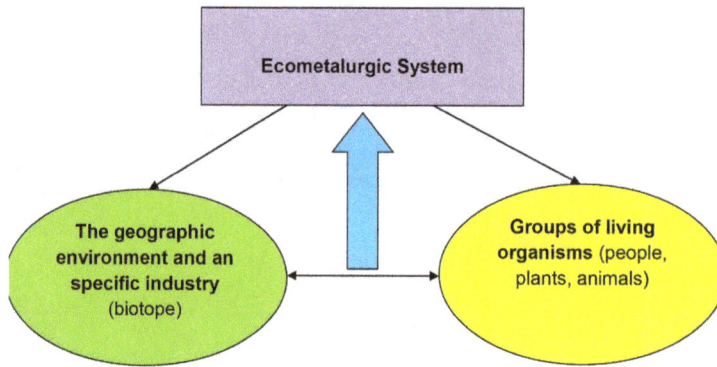

Figure 4. Scheme of ecometallurgic system (ES). Source: own research.

3. The Classification and Characterization of the Ecometallurgic Systems (ES)

We define and characterize two types of ecometallurgic systems (ES), namely:

a.   Monovariabile SE (SEMo)

b.   Multivariabile SE (SEMu)

3.1. The characterization

a.   Monovariable SE (SEMo)—it is characterized by a single magnitude input ($u(t)$; $m(t)$), a single magnitude output ($y(t)$); and a single magnitude of perturbation ($p(t)$).

In **Figure 5**, we present a schematic diagram of an monovariabile ecometallurgic system (SEMO).

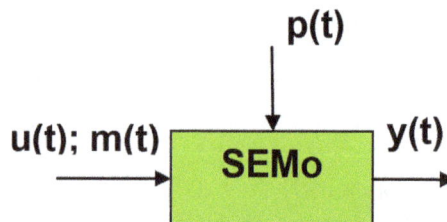

Figure 5. The schematic diagram of an monovariable ecometallurgic system (SEMo). Source: [5]. $u(t)$; $m(t)$—magnitude input; $y(t)$—magnitude output; and $p(t)$—magnitude of perturbation.

SE Monovariable—it is used to study mathematical modeling and simulation. Given the complexity of the metallurgy, simulation and mathematical modeling can have a great importance.

a.   SE Multivariable (SEMu)—it is characterized by several dataset input quantities ($\sum u, m_{i(t)}$), more outputs—the set of output quantities ($\sum y_{j(t)}$), and several sizes of disturbance—the set of disturbance sizes ($\sum p_{k(t)}$).

Figure 6 shows a schematic diagram of ecometallurgic multivariable system (SEMu).

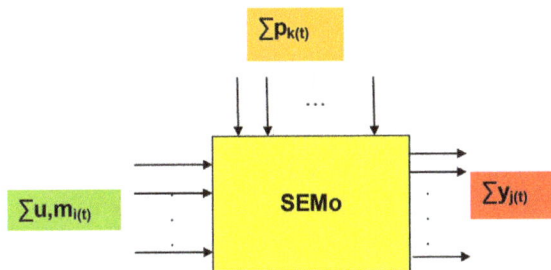

Figure 6. The schematic diagram ecometallurgic multivariable system (SEMu). Source: [5]. $\sum u, m_{i(t)}$—the set of input quantities; $\sum y_{j(t)}$—the set of output quantities; and $\sum p_{k(t)}$—the set of disturbance sizes.

4. About the ecological balance

The concept of ecological balance was among the first to go beyond theoretical scientific studies, becoming an emblematic concept of harmony in the environment.

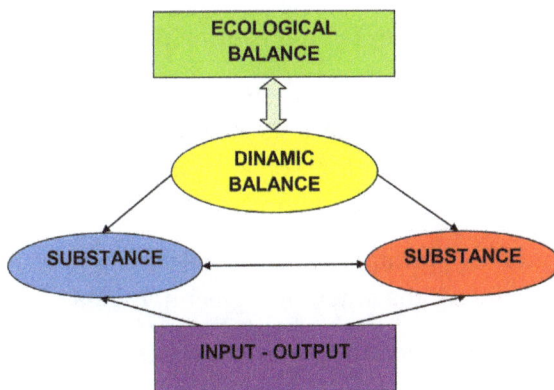

Figure 7. The scheme of the concept of ecological balance. Source: own research.

Ecological balance is a state (an ecosystem) maintained through complex interactions, which aroused particular interest deceleration in terms of theoretical debates and empirical observations.

The study of mechanisms that ensures this status allows the forecast *ecosystem* responses to disturbance anthropic. In terms of exchange of substance and energy, ecological balance expresses *the dynamic balance* ratio between input and output unit.

Figure 7 shows an illustration of the concept of ecological balance.

The maintaining of the ecological balance requires a self as finite nature of resources or space and a virtually unlimited potential of biological breeding populations.

The solutions for maintaining the ecological balance are the recycling and control of the growth which in the ecosystem leads to differentiation of functions for each population, followed by the creation of interdependence and organization of a self-regulating cybernetic system.

5. The principles for the environmental legislation applicable to ensure ecological balance

Among these principles, we consider useful to remember the following:

- The principles of the environmental legislation internally:

- The principle that environmental protection should be an essential element of economic and social policy of the state

- The principle of preventing environmental risks and damage occurrence — this principle has as its main goal to minimize environmental risks, including those from the effect of greenhouse gases.

- The principle of health priority compared with other purposes for use of natural resources — according to this principle greenhouse gases have a major harmful effect upon health.

- The precautionary principle in decision making — any decision, both in technology and in the environmental domain, should rely on this principle of precaution.

- The principle of prevention, reduction, and integrated pollution control — prevention must prevail in any ecological action; in the field of greenhouse gas many decisions of the European Union, which are binding for România as well, predict the slow and gradual decrease in quantity of the emissions and immissions of these gases available before 2030.

- The principle of retention of pollutants at source — the application of this principle is of particular importance to reduce the amount of greenhouse gas emissions.

- The principle of public participation in the environmental protection and improvement.

- The principle of conservation of biodiversity and of ecosystems specific to the natural biogeographical.

- The polluter pays principle — the principle of liability and polluter; this principle, by its punitive character has positive effects including a significant reduction in the quantity of greenhouse gases.

- The principle of state liability and accountability — this principle warns the authorities agaist possible penalties, including the noncompliance requirements on greenhouse gas emissions.

- The principle of sustainable use of the natural resources — the application of this principle implicitely leads to significant reductions in the quantity of greenhouse gases.

- The principle of cooperation in the context of relationship between state, society, and the environment user.

- The principle of the development of international cooperation for environmental protection —this principle takes into account the fact that greenhouse gases know no borders, so from this point of view, international cooperation in environmental protection becomes crucial.

- The principle of integrating environmental policy into other sectoral policies.

- The principles of external environmental legislation.

- The principles of "sic utere tuo".

- The principle of good neighborliness—the application of this principle has direct positive effects in default protection of neighboring countries, including in the area of greenhouse gas emissions.

- The principle of protecting the common heritage of mankind—significant reduction in the quantity of greenhouse gases provides the best conditions to accomplish this principle.

- The principle prohibiting pollution—this principle puts in the foreground the significant reduction of greenhouse emissions and imissions gas.

- The principle of protecting natural resources and common areas—greenhouse gases through their effect contradict this principle; consequently, the accomplishment of this principle implies a significant reduction on the amount of these gases.

These principles are particularly important for ensuring ecological balance. Unfortunately, we must recognize that their application is deficient and therefore their effectiveness remains largely theoretical.

6. Analysis of ecological electric arc furnace (EAF)

Electric arc furnaces are large generators of emissions, with a strong impact on the environment. The main emissions are as follows:

- Powders (powders) resulted during loading operations of raw materials, smelting, refining, alloying, evacuation steel containing heavy metals (Cr, Ni, Zn, Pb, etc.) that may reach values exceeding 15 kg/t steel.

- Process gases smelting and refining, containing mainly CO, CO_2, SO_x, and NO_x.

Of the total dust emissions, 90% are generated during smelting and refining operations. These powders are rich in oxides of iron, manganese, silicon, and aluminum and heavy metals such as nickel, chromium, cadmium, lead, and copper. But their chemical composition is highly variable, being directly influenced by

- composition of raw materials that make up the load EAF;

- melting driving mode;

- refining process used (oxygen gas or ore);

- during smelting and refining processes;

- steel grade that are elaborated.

Table 1 gives the range of variation of the chemical composition of the dust generated during the production of steel in electric arc furnaces in the United States and Germany, the load entirely made up of scrap.

No.	Component	Variation limits (%)		
		USA	Germany	
			Nonalloy steel	Alloy steel
1.	Fe_{total}	16.4–38.6	21.6–43.6	35.3
2.	Si	0.9–4.2	0.9–1.7	17.0
3.	Al	0.5–6.9	0.1–1.5	[a]
4.	Ca	2.6–15.7	6.6–14.5	0.4
5.	Mg	1.2–9.0	1.0–4.5	1.2
6.	Mn	2.3–9.3	0.9–4.8	2.0
7.	P	0.0–1.0	0.1–0.5	[a]
8.	S	0.0–1.0	0.3–1.1	0.1
9.	Zn	0.0–35.3	5.8–26.2	1.4
10.	Cr	0.0–8.2	0.0–0.1	13.4
11.	Ni	0.0–2.4	[a]	0.1
12.	Pb	0.0–3.7	1.3–5.0	0.4

[a]Lack of data. Source: [12]

Table 1. Chemical composition of EAF dust emissions.

In terms of the pollution decreasing, the crucial issue in the electric arc furnace is improving the collection of dust from the process gases both in the oven and work area for improved working conditions in those areas and to respect the limits imposed by legislation labor safety and environmental protection.

Determinants of the above requirements along with increased performance CAE, involves the following:

- expanding gas collection;

- increasing the separation or reduction of the dust content in the gas;

- reducing operation costs by reducing specific energy consumption;

- reducing maintenance costs and investment costs;

- protection against noise;

- improving working conditions.

To stop emissions from falling into the halls' working atmosphere and environment, electric arc furnaces had to be equipped with efficient capture and treatment.

This was also imposed by severe laws in many countries, on breakpoints dust, as shown in **Table 2**.

Country	France	Germany	Norway	Spain	Denmark
Allowable dust limit value (mg/m^3_N)	10	20	25	50	2–5

Source: [12].

Table 2. Limit values for permissible concentrations of dust.

No.	Emission type	Technological phase of the processing	Emission percentage (%)
1.	Primary	Melting	93
2.	Secondary	Loading	2.75
		Evacuation	3.5
		By leaks (door, bowl—vaulted space around the electrodes)	0.75
	Total	Batch duration	100

Source: [12].

Table 3. Weight classification and dust emissions at CAE.

Emission of dust generated during the technological stages of a batch is divided into primary and secondary emissions in the order of their weight in the total amount of dust generated throughout the batch (see **Table 3**).

No.	Heavy metals	Steel type			
		Carbon steel		Inox steel	
		Offset variation (g/t)	Recommanded value (g/t)	Offset variation (g/t)	Recommanded value (g/t)
1.	As	0.06–0.14	0.1	0.01–0.02	0.015
2.	Cd	0.05–1.5	0.25	0.05–0.09	0.07
3.	Cr	0.3–2.0	1.0	12–18	15
4.	Cu	0.3–1.0	0.8	0.3–0.7	0.15
5.	Hg	—	0.15	—	0.15
6.	Ni	0.1–0.6	0.25	3–6	5
7.	Pb	5–20	14	1–3	2.5
8.	Be	—	0.05	—	0.05
9.	Zr	20–90	50	4–9	6

Source: [12].

Table 4. Emission factors for heavy metals in developing the CAE.

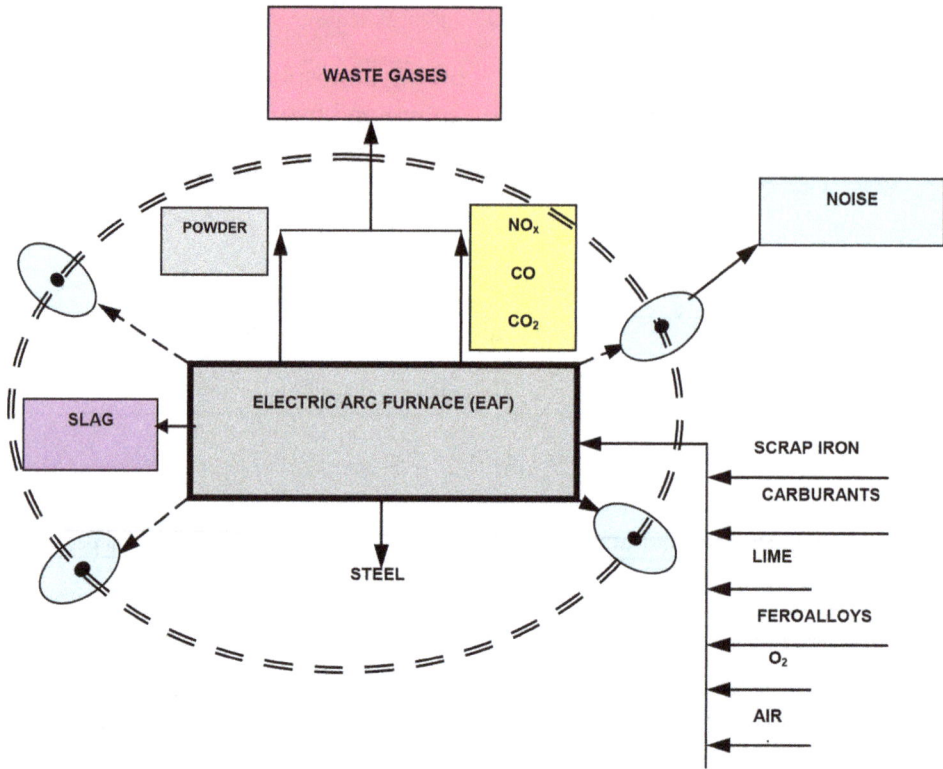

Figure 8. Scheme system of environmental pollution through EAF. Source: own research.

Gaseous phase of emissions that are emitted from the furnace is not only mainly composed of components: CO, CO_2, NO_x, and SO_x, but it also contains other very toxic ones, such as volatile organic compounds (dioxin and derivatives chlorinated benzene and phenol) resulting from burning organic oils that pollute the raw material.

Emission factors in the development of heavy metals in the arc furnace oscillate in a broad difference of values, recommending ATMOS PARCOM work for Europe values shown in **Table 4**.

In **Figure 8**, we present the main scheme of the environmental pollution system through CAE.

For dedusting flue gas discharged from the EAF, it is necessary to perform successively two categories of processes;

• capturing flue gas;

• dedusting flue gas itself.

Capturing the flue gas can be achieved mainly by

• Hoods;

• Suction canopy (the fourth hole in the roof of the furnace);

- Mixed (hood + the fourth hole in the ceiling).

Flue gas dust removal system can be:

- Moist by gas scrubbing;

- Centrifugal cyclone;

- Type filters by using filter bags (textile) or electrostatic precipitators.

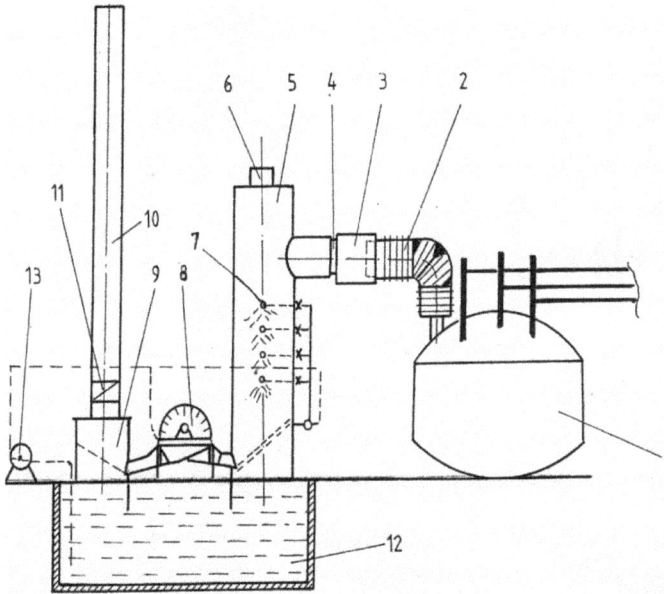

Figure 9. Cyclone wet (electric arc furnace 10 t). Source: [12]. 1—electric arc furnace; 2—suction; 3—mobile sleeve; 4—slot; 5—cooler; 6—tubing safety; 7—spray nozzles; 8—radial disintegrant; 9—separator; 10—cart; 11—throttle; 12—pool; and 13—pump.

The decision on the type of process and the type of facility used for dedusting flue gas discharged from the electric arc furnace is taken mainly based on the following criteria:

- to not adversely affect the process;

- the possibility of grouping the available space;

- keeping a smooth environment;

- operational safety;

- minimum investment volume;

- minimum operating cost;

- capitalization of substances treated.

An example of a wet cyclone used in an electric arc furnace of 10 t is shown in **Figure 9**.

The solution was gas suction through a fourth hole in the roof, proving to be the best way of capturing an electric arc furnace gas.

The suction pipe [2] provided with cooling fins was fixed by the metal construction of the vault of the oven so as to be able to follow all the movements of the tilting and swinging thereof.

Between suction and fixed air purifying, there is a mobile sleeve [3] and a space (gap) [4] necessary both for taking thermal expansion and for regulating the flow of cold air sucked.

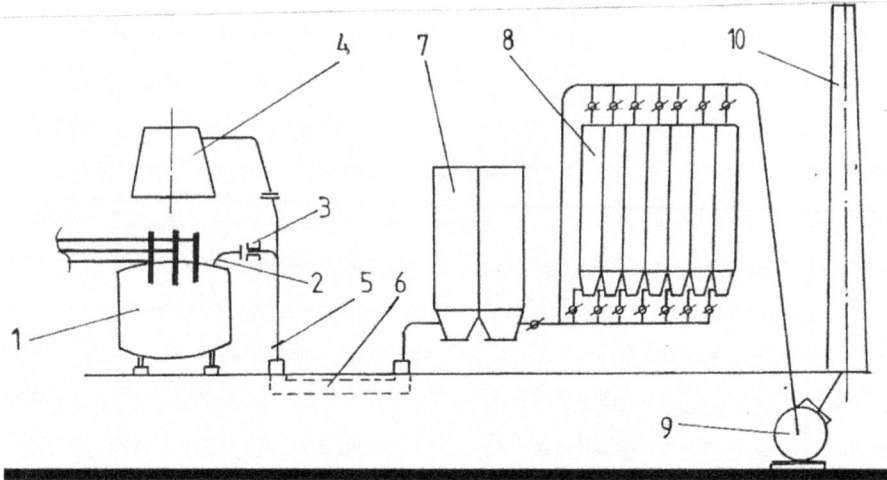

Figure 10. Scheme cyclones exhaust gases from the electric arc furnace (EAF). Source: [12]. 1—electric arc furnace; 2—suction; 3—chamber; 4—mobile hood; 5—keyboards; 6—underground channel; 7—cooler; 8—battery filters; 9—turbo-fan (common); and 10—cart.

The burned gases are cooled entirely up to their dew point in the cooler [5] by spraying water through four nozzles [7]. The cover of the cooler is equipped with a safety pipe [6] for additional entry of air.

The Radial disintegrator [8] is arranged downstream of the gas cooler and extracts therefrom, acting as a suction fan, where a fine treatment takes place at the same time. The gases then enter tangentially into a water separator [9] and are discharged into the atmosphere through a stack [10].

The wash water is recycled to the cyclone reactor. From a pool of water [12] 18 m^3, various points of use are fed by a pump.

The dedusting process of exhaust gases from electric arc furnaces (IPROMET solution) envisages:

• mixed solution for collecting the gaseous phase, both by the fourth hole in Olt (provided with a fitting cooled) and through a mobile hood over the furnace and electric drive (for secondary emissions capture);

• an air cooling gas at ambient temperature;

- filter element—bag filter;

- depression necessary to ensure—through an exhaust chamber (both for the fourth hole in the vault and the vault).

Figure 10 presents the scheme of dedusting plant exhaust gases from the electric arc furnace, used in Romanian steelworks.

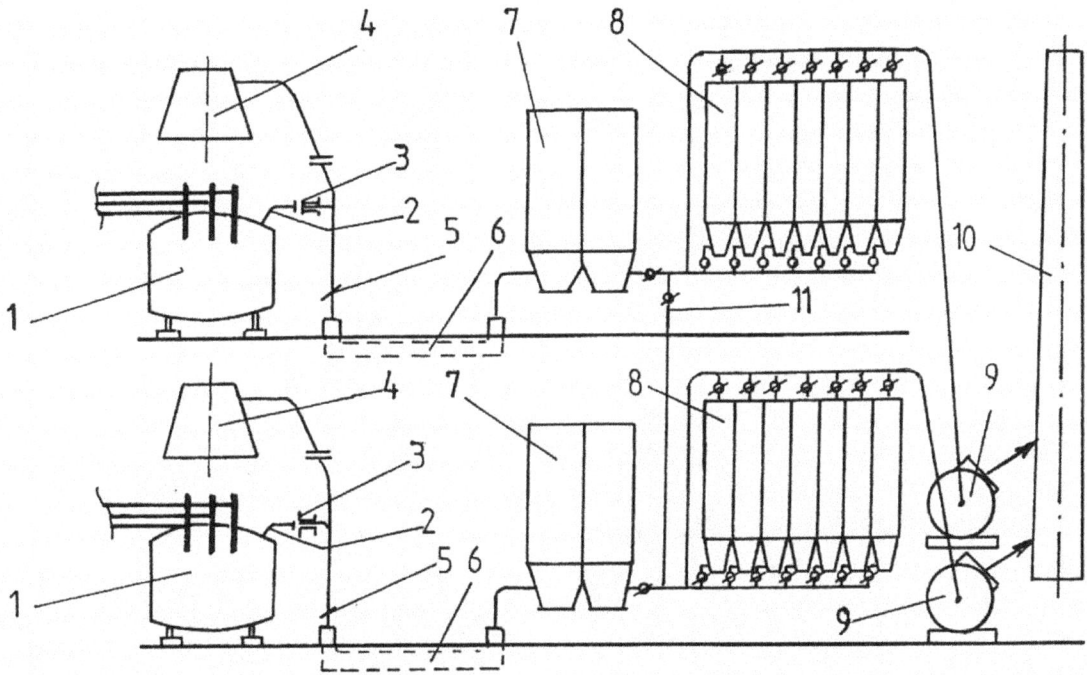

Figure 11. The system of scrubbers in parallel. Source: [12]. 1—electric arc furnace; 2—suction; 3—chamber; 4—mobile hood; 5—flap; 6—underground channel; 7—cooler; 8—battery filters; 9—exhaust; 10—cart; and 11—valve switching.

The flue gases collected through both of the fourth hole in the roof of the oven and the suction pipe (2), gas ceding their heat of reaction in the combustion chamber (3) and through the hood furniture (4). They are directed to an adjustable flap pressure (5) underground channel (6). From this channel, the gases are cooled in cooler (7) and then filtered through the filters battery (8) with bag filters (fabric).

The depression necessary to collect and circulate the gas is ensured by the suction blower (9) and the output bin (10), and the gases are directed after being dedusted.

Aspirations of false air (both by adjustable gap upstream of the combustion chamber and the other leg) directly influence the efficiency of the furnace exhaust gas capture (increased false airflow aspirated gas flow mitigates captured).

Treatment plant may be individual (for each furnace) or in parallel (coupled two by two), each serving one furnace, as shown in **Figure 11**.

Through the throttle switch (11), one can reverse the serviced furnace or that cyclone operation. The coupling system of the plants has the advantage of using a single cyclones for the two furnaces (not simultaneously) so that during repair (revision) of the installation, one of the two furnaces can be operated by the cyclone operation.

Using cyclone influences the regime of the pressure in the oven. Correlated to the increase in false sucked airflow (and implicitely exhaust gas discharged from the oven) caused by the wear dome oven, this requires the use of vaults and cooled walls.

Intensifying the thermal oven and its best possible sealing are goals that lead both to the increase of productivity oven and to reducing specific energy consumption, and they should be made to avoid the risk of uncontrolled ignition of the gas phase route cyclones. To this end, the introduction of the combustion chamber has a decisive role.

In the case of dusting with electrical filters, the gas passes through the electrofilter chamber where deposition electrodes, linked to the ground, and emission electrodes are placed. Due to the difference of voltage of about 75–100 kV between emission electrodes of negative polarity and deposition electrodes, of positive polarity, an electrostatic field is formed.

In the vicinity of the emission electrode, a strong failure of potential is established, which produces the ionization of the gas in this area. Positive ions remain on the emission electrode and the electrons move to the deposition electrodes. on their way The electrons meet gas molecules and dust particles which they ionize negatively. These ionized particles adhere to the deposition electrodes they meet.

The layer of powder deposited can reach a thickness up to 10 mm. It is removed by shaking the deposition electrodeswith the aid of a striking hammer device. Dust collection is achieved in a specially arranged bunker at the bottom of the electrostatic precipitator.

In electric filters, continuous current is used so that the ionized particles travel only in one direction (toward the deposition electrodes).

7. Pollution prevention through afterburner

As shown, after thermal metallurgical processes gaseous combustible substances such as CO, H_2, and CH_4 result. It is proposed that these gases be used, after leaving the contour energy as substitutes for other aggregates of expensive or deficient fuels.

Lately, to increase the efficiency of enthalpy and chemical potential (thermal effect of oxidation reactions—burning) of burnt gas, one need to burn combustible components in the working unit of the aggregate.

This process of modernization, applied, for example, to oxygen converters and electric arc furnace (EAF) is called postcombustion. Since the consumption of CO takes place inside, the method is also considered a way of reducing pollution.

Essentially, the method involves the recovery, even in the technological outline, of the heat of exothermic combustion reaction of CO with oxygen, blown into the workspace via a lance especially designed for this purpose:

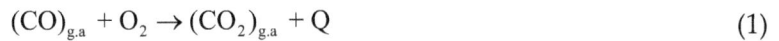

$$(CO)_{g.a} + O_2 \rightarrow (CO_2)_{g.a} + Q \qquad (1)$$

The process efficiency is assessed by postcombustion indication rate, defined as the ratio indicator:

$$\eta_{p.c} = \frac{(\%CO_2)}{(\%CO + \%CO_2)} \qquad (2)$$

Detailed analysis of postcombustion process shows that there are still reservations regarding the technical possibilities for improvement and contributions to the development of theoretical knowledge underpinning the process.

Thus, the materials published so far have failed a systematized existing information. For this reason, the authors of this paper, proposes the following classification of postcombustion processes.

a) Natural postcombustion, in which extra energy is built on the combustion components (CO and H_2), naturally eliminated from the process; combustion occurs upon contact with the jet of oxygen blown into the furnace. This process has two options:

(a.1) Natural free postcombustion based on the furnace burning combustible gases from process gases in the presence of oxygen jet blew right through the walls of the unit, and depending on the placing of the jet, we identify two technologies:

• Natural free postcombustion with nonimmersed jet or, in short, postcombustion nonimmersed jet where the postcombustion is produced in the white space of the melt-existing fireplace.

• Natural free postcombustion with immerged jet or, in short, postcombustion immersed jet, in which case the oxygen jet pierces the layer of slag, producing foaming slag, which is why the process is also found under the postcombustion foamed slag.

(a.2) Forced natural postcombustion performed when the fireplace blows a jet of supplemental oxygen crossing metal melt and slag;

b) Artificial postcombustion that involves blowing of a coal jet and a jet of oxygen at the same time. In this case, postcombustion also involves burning coal and related processes.

c) Combined postcombustion, which involves a combination of the above.

In specific literature, postcombustion is analyzed as the process that occurs in conjunction with other measures (oxy-combustion, foamed slag, etc.). Therefore the authors consider it necessary to distinguish between:

- pure postcombustion, which means postcombustion in a classic oven without other measures;

- pseudo-postcombustion, where postcombustion relates to other modernizing processes.

Since in the case of CAE, there may be hydrogen gas, H_2 coming from the combustion of hydrocarbons added in the combustion process or waste scrap, and it is possible to have a postcombustion reaction:

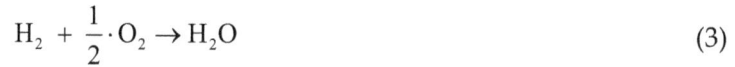

$$H_2 + \frac{1}{2} \cdot O_2 \rightarrow H_2O \tag{3}$$

In these circumstances, we propose that for the calculation of efficiency of postcombustion to use a new relationship that will characterize the complete process:

$$\eta_{p.c.c} = \frac{(\%CO_2 + H_2O)}{(\%CO_2 + \%H_2O + \%H_2 + \%CO)} \tag{4}$$

Based on the general theory of thermo metallurgical installations, CAE part of, we know that when combustible substances (such as CO, H_2, and CH_4) are burnt with flame, a process of dissociation of products of combustion simultaneously occurs, according to the reactions :

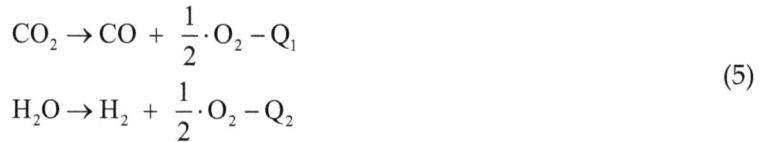

$$CO_2 \rightarrow CO + \frac{1}{2} \cdot O_2 - Q_1$$

$$H_2O \rightarrow H_2 + \frac{1}{2} \cdot O_2 - Q_2 \tag{5}$$

This phenomenon makes a distinction between combustion calorimetry temperature given by the equation:

$$t_k = \frac{H_i}{v_{g.a} c_{p_{g.a}}} \tag{6}$$

where H_i is the calorific value of the fuel [J/kg; J/m³$_N$]; vg.a—the amount of gas flared [m³$_N$ g.a/ m³$_N$; kg] and—specific heat of flue gas [J/m³$_N$] and theoretical combustion temperature:

$$t_t = \frac{H_i - Q_{dis}}{v_{g.a} c_{p_{g.a}}} \tag{7}$$

where Q_{dis} is the amount of heat consumed for the dissociation of CO_2 and H_2O [J/kg; J/m³$_N$]. Theoretical calculations, confirmed by the experiment, show that this lost heat can have values $Q_{dis} = (2...4 \%)H_i$.

At the same time, postcombustion products CO_2 and H_2O can react with carbon and iron in molten metal or through oxidation with the iron in the charge:

$$C + CO_2 \rightarrow 2CO \quad \text{(endothermic reaction)}$$
$$C + H_2O \rightarrow CO + H_2$$
$$CO_2 + Fe \rightarrow CO + FeO \tag{8}$$
$$H_2O + Fe \rightarrow H_2 + FeO$$

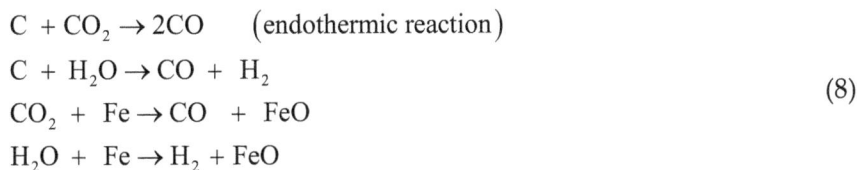

The last two observations lead to the conclusion that at the same time with the postcombustion processes there occur processes of endothermic consumption of the products CO_2 and H_2O. Based on this affirmation, in this paper, we propose that we generally call such a phenomena anticombustion.

Theoretical and experimental study in postcombustion shows that so far not enough consideration has been given to intensify postcombustion processes. One of the theoretical and practical possibilities of intensifying launched by the authors in the research is the postcombustion in ultrasonic field called postcombustion ultrasonic (PCU). This new method assumes that the jets of fluids (e.g., oxygen) blown into postcombustion zones be ultrasonic energy carriers, based on which the processes of mass and heat transfer in the mentioned area to be enhanced.

8. Conclusion

The metallurgical environmental complexity and therefore the Metallurgical Process Technology (MPT) is grounded and by which it interacts systems: The system-Energy-Recycling The Environment (ERE), Ecological system (ECO), Recycling, Reclamation the system (REC-REV).

The concepts of sustainable development (SD) and the total quality (TC) are of particular importance in analyzing correlations between the System Technology Process Metallurgy (MTP) and the other systems.

The Metallurgical Ecosystem analysis has as a starting point ecometallurgic monovariable system (SEMo). This system applies only theoretically, and it is very important for modeling and simulation environment related to metallurgical processes.

The ecological balance is a concept very complex and very difficult.

The especially self-regulating mechanisms and applying to the concept of sustainable development is very important for ensuring ecological balance.

The principles of environmental legislation were also of particular importance for achieving ecological balance. Among these principles we mention: the principle of preventing environmental risk and damage, the principle of priority health compared with other purposes for use of natural resources, the principle of prevention, reduction, and integrated pollution control, the principle of retention of pollutants at source, the principle of public participation in the protection and improving the environment.

Dedusting the flue gas discharged from the electric arc furnace (EAF) has a special significance for its hipopolluting functioning. The main categories of processes to achieve this are flue gas capture and dedusting actual flue gas.

From a wide range of machinery and equipment specific to this field, having as starting point the scheme of system environmental pollution through EAF, in this first part of our article, we presented the cyclone of wet and dry dedusting plant.

The technological development of steel in electric arc furnaces (EAF) is one that is ecologically impaired. The emissions and immissions resulting from this technological process are many and in significant amounts. In conclusion, special care is required from production managers (and not only) to ensure hipopolluting operation conditions of EAF. This concern should begin in the early stages of research both in technology development and designing this complex aggregate.

Achievements in greening the operation of the electric arc furnace (EAF) to develop steels, are relatively modest on a national level.

The costs for installation and commissioning of the capture and treatment of this complex aggregate emissions are significant. Even so, the restrictive environmental regulations in the field constantly force the user to take technological measures to ensure the functioning of hipopolluting EAF.

From this point of view, the specialists in the field should pay far greater attention and importance of scientific research and design.

Author details

Adrian Ioana* and Augustin Semenescu

*Address all correspondence to: adyioana@gmail.com

University Politehnica of Bucharest, Bucharest, Romania

References

[1] Brimacombe, L., Jackson, C., Schofield, N., Artificial intelligence expert systems for steelworks pollution control, La Revue de Métallurgie-CIT Janvier 2001, pp. 111–116, 2001.

[2] Bardet, I., Desmonts, T., Ryckelynck, F., Bourrier, Ph., Video monitoring of visible atmospheric emissions: from a manual device to a new fully automatic detection and classification device, La Revue de Métallurgie-CIT 2000, pp. 1223–1234, 2001.

[3] Ioana, A., Metallurgy's Impact on Public Health, Review of Research and Social Intervention, 43/2013, ISSN: 1583-3410, eISSN: 1584-5397, (ISI-Web of Social Science/ Social Science Citation Index Expanded), Accession Number: WOS: 000328004800011, IDS Number: 266DY, pp. 169–179.

[4] Ioana, A., Semenescu, A., Technological, economic, and environmental optimization of aluminum recycling, Journal of the Minerals, Metals & Materials Society, JOM, 65, 8 (2013), ISSN: 1047-4838 (ISI-Web of Science/Science Citation Index Expanded), Accession Number: WOS: 000322136400007, DOI: 10.1007/s11837-013-0664-6, IDS Number: 187RN, pp. 951–957.

[5] Ioana, A., Elemente de Automatizare Complexă a Sistemelor Ecometalurgice (ACSE) şi de Robotizare, Editura Printech, ISBN: 978-606-23-0246-7, Bucureşti, 2014.

[6] Nicolae, A., Scorţea, C., Lepădatu, Gh., Sisteme ERE (environment-recycling-energy) în industria siderurgică, Editura Fundaţia Metalurgia Română, Bucureşti, 1997.

[7] Ioana, A., Semenescu, A., Preda, C.F., Knowledge management innovation for sustainable development in the context of the economic crisis, WSEAS ISI Proceedings of the 2013 International Conference on Environment, Energy, Ecosystems and Development (EEEAD 2013), Venice, Italy, September 28–30, 2013, pp. 21–26, ISBN: 978-1-61804-211-8.

[8] Ioana, A., Mirea, V., Bălescu, C., Analysis of service quality management in the materials industry using the BCG matrix method, Amfiteatru Economic Review, XI, 26, 2009, pp. 270–276, [ISSN: 1582-9146, ISI-Web of Science/Science Citation Index Expanded], Bucureşti, 2009. Accession Number: WOS: 000267351800004, IDS Number: 462KQ, Research Areas: Business & Economics Web of Science Categories: Economics, Publisher Editura ASE, Piata Romana, Cited References in Web of Science Core Collection: 8,Times Cited in Web of Science Core Collection: 2.

[9] Ioana, A., Bălescu, C., Environmental study of the formation of evacuated burnt gases from a steels making plant, Revista De Chimie 5/2009, pp. 468–471, [ISSN 0034-7752, ISI-Web of Science/Science Citation Index Expanded], Bucureşti, 2009. Accession Number: WOS: 000267459400008, IDS Number: 463VB, Research Areas: Chemistry; Engineering Web of Science Categories: Chemistry, Multidisciplinary Engineering, Chemical, Publisher Chiminform Data SA, Bucureşti, Cited References in Web of Science Core Collection: 10.

[10] Ioana, A., Semenescu, A., Preda, C.F., Elements of best management for metallurgical technological plants, Metalurgia International 18(1/2013), ISSN 1582-2214, (ISI-Web of Science/Science Citation Index Expanded), Bucureşti, 2013, pp. 165–167. Accession Number: WOS: 000315368200037, IDS Number: 095XL, Research Areas: Metallurgy & Metallurgical Engineering, Web of Science Categories:Metallurgy & Metallurgical Engineering, Publisher EDITURA ŞTIINŢIFICĂ FMR, Bucureşti, Cited References in Web of Science Core Collection: 9.

[11] Ioana, A., Semenescu, A., Preda, C.F., Metallurgical marketing mix (MMM) elements, Metalurgia International, 18(1/2013), ISSN 1582-2214, (ISI-Web of Science/Science

Citation Index Expanded), Bucureşti, 2013, pp. 156–159. Accession Number: WOS: 000315368200035, IDS Number: 095XL, Research Areas: Metallurgy & Metallurgical Engineering, Web of Science Categories: Metallurgy & Metallurgical Engineering, Publisher EDITURA ŞTIINŢIFICĂ FMR, Bucureşti, Cited References in Web of Science Core Collection: 10.

[12] Ioana, A., Nicolae, A., Predescu, Cr., Sandu, I.F., Sohaciu, M., Calea, G.G., Conducerea Optimală a Cuptoarelor cu Arc Electric, Editura Fair Partners, Bucureşti, ISBN 973-8470-04-8, 2002.

7

Life Cycle Assessment Method – Tool for Evaluation of Greenhouse Gases Emissions from Agriculture and Food Processing

Zuzana Jelínková, Jan Moudrý Jr, Jan Moudrý, Marek Kopecký and Jaroslav Bernas

Additional information is available at the end of the chapter

Abstract

The chapter focuses on the use of the Life Cycle Assessment method to monitor the emission load of foods from different systems of farming production. The products of the conventional and organic farming production intended for public catering are compared within the SUKI and UMBESA international projects. Conventional farming is mainly characterized by high inputs of mineral fertilizers, chemical pesticides, the use of hormones and stimulants in animal husbandry. It is a system based on the highest possible yields without respecting the natural principles of nature. Conversely, organic farming is a system of production established by the legislation that respects fundamental natural cycles, such as crop rotation, ensures welfare of animals, prohibits the use of fertilizers, pesticides, and other substances of synthetic origin. However, lower yields are a big disadvantage. In the Czech Republic, only about one tenth of the agricultural fund is currently used for organic farming. Arable land constitutes only about 10% of the total area of agricultural land, other areas are mainly grasslands and orchards. The work primarily aims to answer to the question whether the selection of foods may contribute to decrease in greenhouse gas emissions, which is a part of the objectives of many policies. Besides the comparison of agricultural production, processed and unprocessed foods, local and imported foods and fresh and stored foods were compared as well. The Life Cycle Assessment (LCA), which is used to assess environmental impacts of products and services throughout their entire life cycle, was used to quantify the emission load. This method may be briefly characterized as a gathering of all inputs and outputs that take place during the production in the interaction with the environment. These inputs and outputs then also determine the impact on the environment. The LCA consists of four successive and iterative phases: This concerns the definition of objectives and scope, inventory analysis, impact assessment and interpretation of the results. The LCA was originally developed for the assessment of impacts of especially industrial products. Certain methodological problems and deficiency, which bring a level of uncertainty of the results, have been caused by its adaptation to agricultural product assessment, but this method is still recommended for comprehensive assessment of environmental impacts of agricultural production and the comparison of different agricultural products. In this study, a Cradle-to-Gate assessment was performed, which means that the impacts of products (in this case the emission formation) were evaluated only to the delivery of foods to public facilities, further treatment and waste management was not assessed. About 20 most frequently used foods for school catering facilities were compared. The results of the project confirm the general assumption about the less emission load of unprocessed, fresh and local products. It may not clearly state that products from organic farming produce less emissions when comparing agricultural systems. It always depends on the particular crop. The absence of synthetic substances such as fertilizers and pesticides reduces the emission load of organic farming, on the other hand, a higher number of mechanical operations and especially the lower income clearly increase the emission burden, therefore, in several cases, lower emission loads of crops were achieved using the conventional farming system. However, less emission may be achieved within the organic farming system. Among 11 evaluated agricultural products, 8 organic products and only 3 conventional ones go better. The situation is different regarding the following phases of food production, processing and transport. The transport phase significantly worsens the environmental profile of organic foods, because transport distances are too far due to insufficient processing capacity and underdeveloped market networks, and often exceed the emission savings from the agricultural phase. On the contrary, conventional foods are carried within relatively short distances, therefore the final emission load of conventional foods is in many cases fewer than the load of organic foods. This fact is also confirmed by the result of the study, because among 22 evaluated foods, organic food goes better in 11 cases and conventional food in 11 cases as well.

Keywords: LCA, conventional farming, organic farming, greenhouse gases, food

---

1. Introduction

Currently, agriculture is one of the largest anthropogenic activities with global impact. The area of agroecosystem that covers about one third of the landmass [1] is directly related to the need of humans to survive and it follows the population growth to a large extent. With the growing population curve, the pressure on natural habitats and their conversion to agricultural land and intensification of farming on existing agricultural land also increases. Since the population growth continues very rapidly and also the consumption of meat, respectively animal products, and the energy consumption in agriculture increase, we cannot expect that in the foreseeable future, a spontaneous reversion of the trend of increasing environmental load will come [2].

The environmental load increase impacts the soil, water, biodiversity and, last but not least, the atmosphere. Climate changes and anthropogenic contribution to them have become a frequently discussed issue in recent years. It is not clear yet to what extent these changes are natural and to what extent they are influenced by human activities. Many questions have not been answered yet and the discussion on whether the climate change is determined by natural evolution or negative consequences of human activity is still held [3]. Just the anthropogenic share of changes, especially in terms of GHG (Greenhouse gases) emission production, may be regulated while this activity is one of the priorities of sustainability.

Climate changes have a significant impact on agricultural systems in the world and can be a crucial factor in ensuring sustainable food production. [4] states that, within the European Union, the largest polluters are energetics, which releases 27.8% of anthropogenic greenhouse gas emissions, transport with 19.5% and industry with 12.7%. Agriculture is with 9.2% in fourth place. Current agricultural trends tending to sustainability should establish more environmentally friendly ways while maintaining the ability of the population food assurance. In order to take steps in this direction, it is necessary to understand agricultural impacts and be able to quantify them. In the case of greenhouse gases, the accurate quantification is quite difficult. However, there are some methods that can help to implement it. One of the methodological tools is the Life Cycle Assessment - LCA. It can be used to quantify GHG emissions, respectively emission saving options. It is a transparent scientific tool [5] which evaluates the environmental impact on the basis of inputs and outputs within the production system [6]. Additionally, LCA analysis currently offers (as one of the few tools) a comprehensive approach to assess the environmental effects [7]. A very valuable tool is LCA analysis thanks to its ability to incorporate and compare different farming systems, their individual processes and products and most of their environmental impacts [8].

Considering the choice of farming system, respectively changes within particular farming systems, as a tool for mitigation, we need to quantify their total impact first and to find the most problematic areas in terms of emissions that can provide space for an effective change. The choice of farming system could be one of the ways to reduce the anthropogenic share of GHG emissions while organic farming seems to be one of the ways. In the last decade, organic farming has become an important element in the environmental friendliness policy and the policy of quality of food in Europe because, inter alia, it reduces the use of synthetic fertilizers and other chemicals such as pesticides [9]. However, mitigation can be achieved also within conventional and integrated farming systems and within food production in general. Reduction of emissions and environmental load in general is a necessary way to long-term sustainability within current population conditions.

2. Literature search

2.1. Climate change and agriculture activities

Anthropogenic activities have a very strong impact also on the environment. With increasing population curve, globalization, technological progress and higher consumer demands, also environmental pressure and environmental impacts grow. There are many impacts from impacts on water, soil, biodiversity to the impacts on the air. Just the anthropogenic air pollution and its relation to climate changes is a big current issue.

Agriculture is ranked among the five major anthropogenic activities contributing most to the production of greenhouse gases. Global GHG emissions from agriculture reach values from 5,1 to 6,1 billions tons of CO_2 equivalent [10]. [11] sets out the share of emissions of greenhouse gases (CO_2, N_2O and CH_4) from particular fields of human activities, while his findings indicate that agriculture in 2000 contributed to the anthropogenic emissions with 13.5%. More than one third of agricultural emissions are field emissions (especially N_2O), methane (CH_4) makes up about one third. Also [12] states that agriculture contributes to the worldwide emission production with the share of 10-12%, while until 2030, we can expect an increase of even half these values [13]. Agriculture is a significant emission producer in the EU also according to [14]. The total share of GHG emissions from agriculture within the EU-27 was 10.1% in 2011 [15]. We can find similar values also in the paper by [16] who states that this share within the EU-15 was 10.2% in 2009. In the Czech Republic, the share of agricultural emissions in total greenhouse gas emissions is calculated at 6.42% [17].

According to [18], 29% of emissions produced within the EU is related to the food production. However, these emissions arising within food production are related not only to the field cycle but also to the production of fertilizers and agrochemicals, processing or all process transport. [18] sets the share of food production to anthropogenic emissions to 22-31% while the most significant proportion (15%) is related to transport.

[19] also stated the high dependence of agriculture on non-renewable materials and to a great extent, the resulting increased GHG emissions production. Agriculture produces emissions in many ways. For example, CO_2 is released during the consumption of fossil fuels or within reduction of organic matter content in the soil. N_2O is released as a result of fertilizer application, CH_4 from the digestive tract of some livestock species. We can conclude that the amount and composition of our diet reflect the specific features of particular technological processes in agriculture and thus the different GHG emission production. Therefore, the change in the way of nutrition in industrialized countries can be extremely important to ensure sustainable development (admittedly conditional on the stabilization of anthropogenic GHG emissions) [20].

2.2. Farming systems

Production systems have their own characteristics and can be categorized into groups e.g. according to density and the resulting impact on the environment. Conventional farming systems are commonly widespread, alternatively, there are integrated and organic farming systems.

2.2.1. Conventional farming

Conventional farming is the most common way of farming in agriculturally advanced countries. Its main objective is to maximize production. Other farming aspects are secondary. Conventional farming is implemented in various intensity degrees. Environmentally friendly processes beyond the ordinary laws are not enforced and monitored. Still, conventional farmers can implement these processes and farm in accordance with environmental protection. However, the European Union introduces a number of rules and legislative provisions for conventional farming leading to limiting inputs in order to protect the environment. On the contrary, in its extremely intensive forms, the conventional farming often leads to excessive environmental damage. The precision farming is a technologically advanced form of conventional farming that reduces environmental load to some extend through more efficient and optimized inputs.

2.2.2. *Integrated farming*

Integrated farming is a kind of an intermediate step between conventional and organic farming systems, originally based on integrated plant protection and extended to other agrotechnical processes. Its objective is the sustainability of farming system and it is largely focused on procedures friendly to the environment. However, unlike organic farming, it is not strictly limited by legislation and it is possible, if necessary, to apply procedures that are forbidden within organic farming (e.g. the use of some agrochemicals).

2.2.3. *Organic farming*

Organic farming is a special kind of farming that cares about the environment and its particular components through restrictions or bans on the use of substances and procedures that burden the environment or increase the risk of contamination of the food chain. Within livestock breeding, it ensures their behavioural and physiological needs in accordance with the requirements of specific legislation. It becomes an environmentally friendly alternative to other farming systems [21]. The main goals of organic farming include:

• Maintenance and improvement of soil fertility.

• Genetic resources protection and biodiversity maintenance.

• Preservation of landscape features and their harmonization.

• Water management, keeping water in landscape and the protection of surface and groundwater against contamination.

• Efficient use of energy, focusing on renewable resources.

• The pursuit for maximum nutrients recirculation and a prevention of the entry of extraneous substance into agroecosystem.

• Production of quality food and raw materials.

• Optimization of life for all organisms, including humans.

Organic farming systems create more potential to reduce greenhouse gas emissions than conventional. The biggest difference is due to the absence of chemical fertilizers. The Farming Systems Trial at Rodale Institute, an American long-term research comparing organic and conventional agriculture, states that the introduction of organic farming nationwide in the USA would manage to reduce CO_2 emissions by up to a quarter due to increased carbon sequestration in soils [22]. The disadvantage of organic farming is less production per the area unit that increases the unit emission load. [23] states that yields of organic farms are on average 17% lower than within conventional farming systems. The impact of organic system on the mitigation is usually measured per the area unit in order to enhance the objectivity. However, it is important to convert it also to the production unit.

2.3. A Life Cycle Assessment

The aim of the assessment of the effects of agricultural products on the environment is to evaluate their impact on environment sustainability [24], especially in terms of food consumption patterns [25]. As stated by [26], the system sustainability can be evaluated on the basis of inputs and outputs and their conversion to CO_2e. [27] states that the measurement of GHG emissions suffers from certain inaccuracy. The reason for this error is that emissions in agriculture are influenced by complex biological processes with a wide range of variables.

There are some suitable methods to assess environmental impacts of agricultural activities [28] such as Life Cycle Assessment (LCA), Ecological Footprint or Emergy Analysis. [29]. The LCA method may be briefly characterized as an assessment of all inputs, outputs and possible impacts on the environment during the entire life cycle [30]. LCA analysis is a tool that enables to assess environmental impacts within the product life cycle. Social or economic aspects may be included as well, however, the calculation of their impacts has only just begun [31] and the main focus is on the environmental component which evaluates, according to [32], the environmental impact of a product based on the assessment of the material and energy flows, that the monitored system shares with its surrounding space (environment).

[33] states that the LCA is an appropriate instrument because it enables to express the relationships between the food production, transport and production of CO_2.

With the LCA analysis, the impact categories - the impact on climate, water pollution and air pollution - are mostly evaluated. Whereas, impacts such as biodiversity or pesticide toxicity are seldom evaluated because of methodological problems [34]. The LCA study consists of four basic stages: Definition of objectives and the scope, Inventory, Impact assessment and Interpretation [32].

2.3.1. Goal and Scope definition

In the first stage of the LCA analysis, it is necessary to define the objective and the scope of the paper before the actual start [35]. The study goal and scope definition determine the next procedure character and the circumstances in which the study outputs are valid [32]. [36] requires to establish a study goal and scope while the study scope means to determine the product system, the functional unit and system boundaries, to determine allocation rules, the assessment methodology, hypothesis and limits and data quality.

In the objectives of the study, there must be clearly specified who it is addressed to, the reasons for the study and the intended use of the results [36]. This increases the transparency of the study and the comprehensibility of the context of the results since different recipients emphasise different aspects.

The study scope results form goals and is determined by financial resources of the ordering authority and the available time of the processor [5]. The study scope describes the most important methodological choices, hypothesis and limits [35] that are described below.

2.3.1.1. Function and functional unit

To compare products (systems), it is necessary to define also the functional unit. The functional unit is described as a quantified performance of a product system which serves as a reference unit in a study of life cycle assessment [36]. It is an essential element which all study results are related to. It must be chosen so as to be easily expressible and measurable. The functional unit is the starting point for searching for alternative ways how to fulfil the function with a lower negative impact on the environment [5]. [37] states that the determination of functional units is as a crucial step especially when comparing systems with different levels of production per hectare such as conventional and organic farming system. [38] sees fit to set the production unit instead of the area unit as a functional unit. On the contrary, [9] recommends to involve both functional units into calculations and perform the calculations for both the unit area and the unit of production. This is confirmed also by [39] who states that LCA analysis outputs are usually set per the production unit. Some authors, such as [40], state that LCA outputs should by calculated in relation to the area unit allowing the better expression of environmental load carrying capacity. With the LCA analysis, we cannot perform both calculation methods and use the production unit as well as the area unit as a functional unit [2].

2.3.1.2. System boundaries

Each product system consists of a variable number of processes involved in the product life cycle. However, the product under consideration is often related to other processes that may no longer be important for the LCA study. The system boundary serves to the separation of essential and non-essential processes of the product life cycle. Since the choice of system boundaries significantly affects LCA study outcomes and in addition, its intensity and complexity, system boundaries should always be well considered and clearly defined. The choice of system boundaries is carried out with regard to the studied processes, studied environmental impacts and selected complexity of the study. Not-including any life cycle stages, processes or data must be logically reasoned and clearly explained [32].

Determination of system boundaries is always a very important step, especially in the area of food production and agriculture, where the clearly identifiable technological processes and systems meet the natural processes and procedures influenced by a number of factors [41]. The system boundary defines which unit processes will be included in the monitored system [36]. The system boundary definition virtually defines which life-cycle stages will be analysed (in the case the whole life cycle was not included) or what unit processes and elementary flows will or will not be considered. The system boundaries can be restricted to the processes within the farm [42], or can extend into other phases from pre-farming processes, through transport and storage, to the end user, respectively consumption. [43] states that although it would be desirable to include the entire product cycle, most studies of food production omit some phases, usually trade and other related sections. Their impact is mostly negligible in relation to e.g. the agricultural phase [44]. When comparing conventional and organic farming systems, we can also omit the calculation of load from buildings and infrastructure because there are only small differences between farming systems while slightly more noticeable difference is apparent within animal production [45].

System boundaries determine not only which processes will be incorporated into the product scheme, but also define the geographic and temporal scope of the study to determine its purview. Defining the geographical scope (local, regional, national, continental or global) or determination of the exact study location is important for the environmental aspects of various material and energy flows because their impacts may be different in different geographical conditions. E.g. due to different ways of development of power in each country, the environmental impact of power development and hence of energy consuming processes is different. Using unsuitable system boundaries or oversight of important factors such as the place and method of energy development can lead to false results.

2.3.1.3. Allocation principles

During the life cycle assessment, the study authors are very often confronted with the fact that the product system has at its end more than one output. In these cases, we use the allocation. Allocation means the assignment of the share of total environmental burden to particular outputs [32]. The Standard recommends to avoid the allocation whenever possible, e.g. by extending systems or sub-division processes [36].

In the case we cannot avoid the allocation using the above mentioned methods, the Standard proposes to use the allocation based on the physical principle such as weight or energy content of final products.

2.3.1.4. Data quality

The quality of data entering the LCA study is to be determined in view of temporal, spatial, technological, data sources (it must be determined whether primary data required or secondary data can be used), their accuracy etc. It concerns the determination of all requirements for the input data [5].

2.3.2. Life Cycle Inventory

The inventory tasks is to collect environmentally important information about relevant processes involved in the product system. Inventory collects information about unit processes at first and subsequently, an inventory of inputs and outputs of the system and its surroundings is carried out. The goal is the identification and quantification of all elementary flows associated with product system. Inventory analysis is the nature of the technical implementation of LCA studies. It is an essential part of a study, has high demands for data availability, practical experience in modelling product systems and, in the case of using database tools, it is necessary to master them perfectly and to understand their function [46]. The inventory phase principle is data collection that is used to quantify values of the elementary flows. This phase represents a major practical part of the LCA study, time consuming and with demands for data availability and author's experience with modelling product system studies [47].

2.3.3. Life Cycle Inventory Assessment

The inventory results should be presented in clear form, how much and what substances from the environment enter the system and how much get out. These results serve for subsequent life cycle impact assessment [48]. The aim of the life cycle impact assessment is to measurably compare the environmental impacts of product systems and to compare their severity with new quantifiable variables identified as impact category. The impact categories are areas of specific environmental problems such as global warming, climate changes, acidification, eutrophication, ecotoxicity and others. Already in the phase of definition of the LCA study scope, it is necessary to describe what impact category will be applied and which of their environmental mechanisms will serve as a basis for impact assessment [46].

2.3.4. Interpretation

The outcome of the LCA study is a large amount of different values from the inventory as well as from the life cycle assessment. An important task for the study author is to sort the data and their appropriate and understandable interpretation [32]. The need for proper interpretation is also stated by [49] who states that on the basis of LCA outcomes, there are often taking steps with significant economic, environmental and other impacts, while there is the risk that incorrect and misleading interpretation of outputs can lead to a deepening of existing or creating new problems. Since the form of presentation of data often affects their meaning, the life cycle interpretation has become an integral part of LCA studies and gained some rules. On the general, interpretation of LCA consists of structuring data with regard to the most important processes or process groups and the most important substances, performing sensitivity analyses and evaluation of the uncertainties of the study, discussion of the data meaningfulness in relation to the study completeness and the input data quality, and the final summary and formulation of realistic recommendations.

3. Goal of the study

The main objective of the Czech - Austrian SUKI (Sustainable Kitchen) project was to assess the total amount of GHG emissions produced by public catering facilities.

These emissions originate both within energy consumption for the kitchen operation (ie. lighting, heating, ventilation, cooling, operating kitchen appliances, cooking process), but mainly in the food production, processing and transport to catering facilities. While direct energy consumption in the kitchen can be determined relatively easily, emissions from food production are unexplored areas in the Czech Republic. The project set the target to answer following questions using the emission quantification:

- What is the influence of the production method (conventional, organic) on the GHG emission production?

- What is the influence of the place of the food origin (region / outside the region) on the GHG emission production?

- What is the influence of the food processing method (raw, processed, fresh, frozen) on the GHG emission production?

By answering these questions, we can deduce the possibilities and limits of greenhouse gas emission savings without compromising the food quality which is also subject to the actual selection of foods, meals and a preparation process. The aim is to promote catering facilities on the path to sustainable production and at the same time to the food nutritional quality improvement. Through targeted food selection, they can take a step towards sustainable development and a healthy diet, contribute indirectly to the global reduction of greenhouse gas emissions while promoting regional organic farming.

4. Methodological procedure

In the first project stage, it is necessary to identify the most widely used ingredients heading for school catering facilities. For this purpose, we used annual lists of purchased raw materials from partner catering facilities that were processed by tabulating and from them, all the ingredients that made up at least 80% of the raw materials used kitchens during the year were selected. These lists also provide a good comparison between Czech and Austrian cuisines.

The second step and the focus of this chapter was to evaluate the emission load of individual foods from the list of most common foods. There was used the simplified Life Cycle Assessment method in which only the Climate change Impact category was assessed. Detailed description of the LCA methodology is shown in the literature review, the following text describes practical method implementation.Food emission load evaluation using the LCA method

4.1. Goal and Scope definition

On the basis of evaluation of consumption of involved catering facilities, 11 most commonly used products were selected. When work them into other raw materials, we can expand the list to final 22 products that heading for school kitchens. For each product, a comparative study focused on the comparison of organic and conventional versions, imports and regional variants was elaborated, if possible, the also a comparison of the fresh and stored product was made, as well as a comparison of different stages of processing. The results should serve as an answer to the question whether the selection of the food contributes to reducing greenhouse gas emissions. The target group are the chiefs of kitchens, school principals, cooks, diners, farmers, suppliers, as well as actors at the regional and national political level.

Evaluated systems were modelled with the cradle to gate principle, thus the product system of particular foods was terminated at the point of entry into the school canteen. The following presentation of food and related activities, as well as waste management of the product and its packaging materials were not included in the LCA. One kg of the final food was selected as a functional unit. In the case the allocation was necessary, the weight-economic allocation was used.

4.2. Life Cycle Inventory

At this stage, it was necessary to collect the relevant data relating to the entire product system. The product system was divided into sub-processes: agriculture, processing and trade. For agriculture, inputs relating to the consumption of seeds, fertilizers, pesticides and fuel within agricultural operations for crop production, feed consumption, energy and fuel within the livestock sector were surveyed. Emissions from nitrogen fertilizer application within crop production calculated according to the methodology [50] and emissions from manure management in the livestock production, calculated according to the methodology [51], were integrated into agriculture. A general framework for crop and livestock products is shown in Figure 1 and Figure 2.

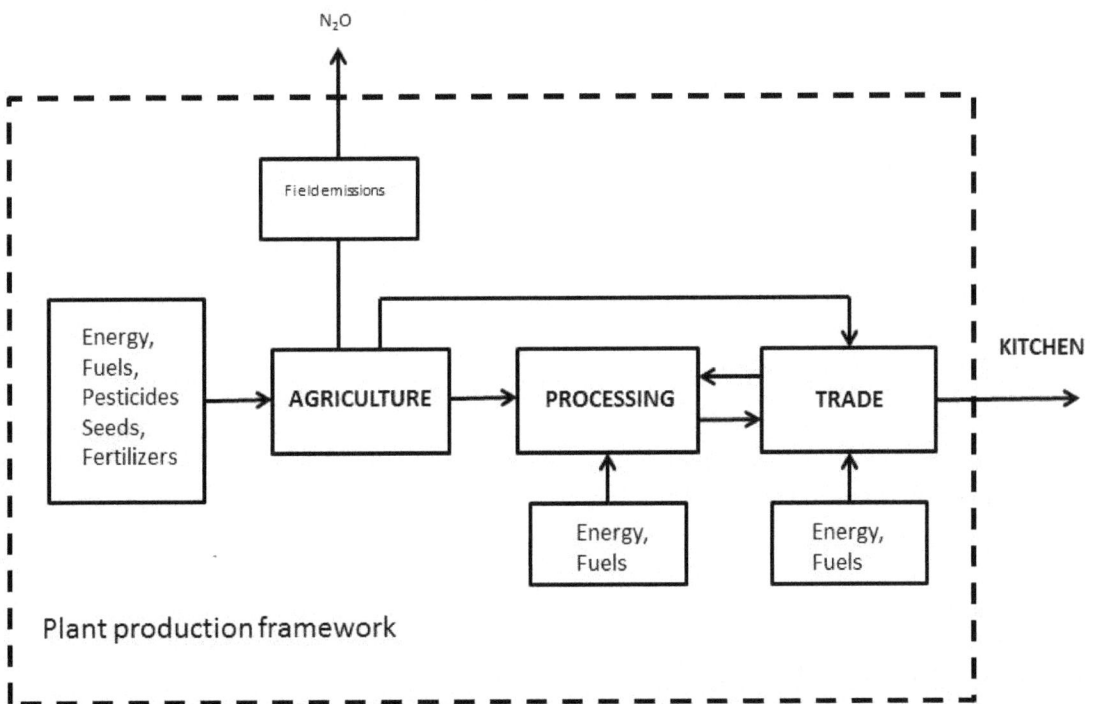

EXCLUDED PROCESSES: water consumption, infrastucture, waste management

Figure 1. Framework of plant food product LCA

For processing, the data on energy consumption were collected, within the trade, it was travel distance, information on cargo and storage time of various foods. All data was obtained primarily from farmers, processors and traders, absent sufficient data, it was supplemented by data from available databases, especially the Ecoinvent database.

From a geographical point of view, regarding the data quality, the data corresponds primarily to the Czech Republic, secondarily to Central Europe. In terms of time, data corresponding to the term 2000 - 2012 were obtained, from a technological point of view, data corresponds to the widely used average technologies.

N$_2$O, CH$_4$

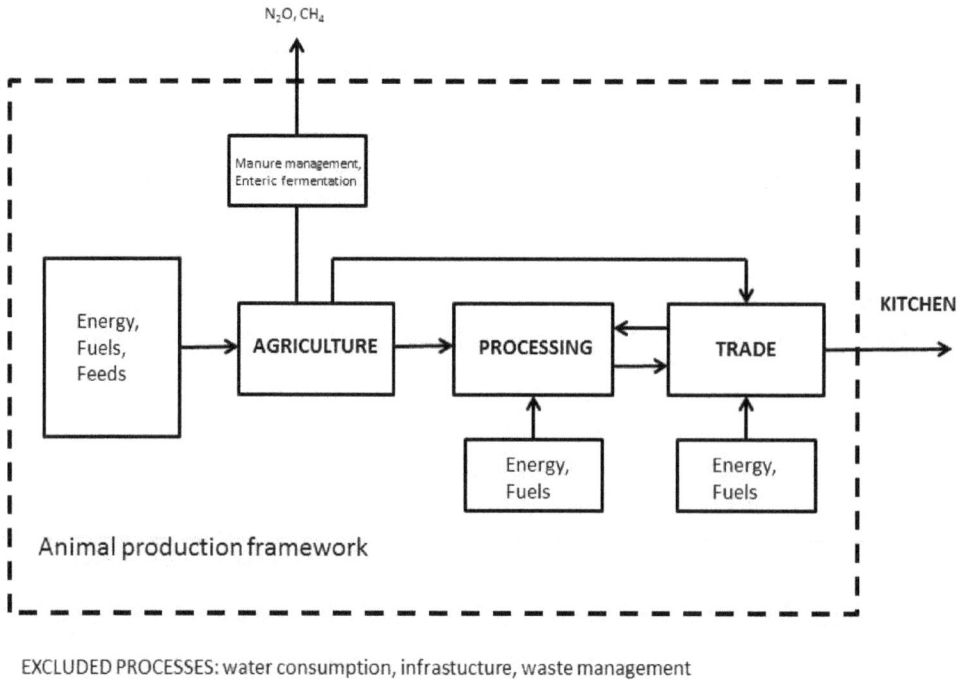

EXCLUDED PROCESSES: water consumption, infrastucture, waste management

Figure 2. Framework of animal food product LCA

4.3. Life cycle inventory assessment

The results were calculated using the SIMA Pro software. To obtain the necessary results, the Recipe Midpoint (H) Europe method has been chosen as a characterization model. Results come from the climate change impact category and they are expressed in kg of a carbon dioxide equivalent (CO$_2$e).

4.4. Interpretation

Result interpretation and discussion is given below.

5. Results

Based on the analysis of the annual consumption of foods of participating catering facilities, there were 22 of the final products which constitute the largest food consumption selected.

5.1. Emission load in food production

5.1.1. Agricultural phase

A basic emission load resulting from agriculture involves the calculation of greenhouse gases in the field phase. In the context of comparing the formation of greenhouse gas emissions in the cultivation of selected crops and breeding of selected species within conventional and organic farming systems, the total greenhouse gas emissions with twelve agricultural products were observed. This total amount sum was divided into subgroups within crop production: agricultural engineering, fertilizers, pesticides, seed and field emission, and in the context of animal production to: feed consumption, manure management, and in the case of cattle on enteric fermentation.

In the case of crop production, the conventional farming system differs from the organic one in the total CO_2e emissions production as well as in the production within subgroups. Although the production of GHG emissions differs within particular subgroups, in total with most studied crops, the production of CO_2e is lower in the organic farming system. In the primary agricultural study, [52] monitored a set of crops including wheat, rye, potatoes, onions, carrots, tomatoes and cabbage, while the higher greenhouse gas emissions expressed as CO2e within the conventional farming system in the Czech Republic were found with all investigated crops except onions. The greatest differences were found with carrots and cabbage where the ecological variants produced almost 60% lower emissions than the conventional variant. The extension study [53] complements the study with the comparison of emission load of organic and conventional apples and rice, where the results showed almost the same burden for rice and in the case of apples, 33% lower emissions within organic farming. Another extension study [54] comparing garlic proves again 40% lower emissions when grown in the organic farming system. In conclusion, it can be summarized that in the context of plant production, eight of ten evaluated products were better as an organic variant, one raw material showed the same emissions in both variants and only one crop was better in the conventional variant. Results and emission savings are summarized in the Table 1.

Group	Product	Organic*	Conventional*	save BIO
corn products	wheat	0,4218	0,4606	8%
	rye	0,2972	0,5364	45%
	rice	0,6197	0,6266	1%
vegetables products	potatoes	0,1256	0,1446	13%
	cabbage	0,0329	0,0774	58%
	carrot	0,0411	0,0987	58%
	tomato	0,0671	0,0871	23%
	onion	0,0997	0,0828	-20%
	garlic	0,2480	0,4306	42%
fruit products	apple	0,0568	0,0848	33%

*in kg CO_2e/kg of products

Table 1. Emission of GHGs from the plant production (agriculture phase only)

Comparative studies show positive and negative factors of organic farming which are mainly lower yields and specific agronomic rules. It coincides e.g. with findings by [55]. The organic farming is more agricultural operations intensive as compared with the conventional one. For most crops, emissions from production of one kg are higher due to more intensive agricultural technology (especially mechanical protection against pathogens), while the difference is even increased by generally lower yields in organic farming. Emission load within the agrotechnical phase in the organic farming system is increased also by some operations related to pre-seeding soil preparation. The possibility of reducing GHG emissions by changes in agricultural technology is highlighted e.g. by [56] who identifies the main potential for reduction within tillage.

The fundamental difference between the conventional and organic farming system in terms of GHG emissions is obvious within fertilization. While organic farming uses organic fertilizers (especially manure or slurry), the use of synthetic fertilizers within the conventional farming system increases significantly the share of emissions. This is stated also by [57] who gives synthetic fertilizer decrease as one of the main tools for reducing CO_2e emissions. From an economic perspective, the nitrogen in organic farms is financially much more demanding than industrially produced nitrogen. This is a powerful incentive to try to prevent losses and learn how to use recycling technology [58]. Timing and management of nitrogen application are crucial. Soil mineralization processes should deliver components to plants when the plants are most in need [10]. In conventional farming, GHG emissions are increased also due to the use of pesticides. In organic farming, this load is completely eliminated, respectively, transferred to the agrotechnical phase in the form of mechanical plant protection. However in total, it is a relatively low proportion of total emissions. [59] can see here another opportunity to save emissions.

Within plant production, in organic farming, there is space for reducing greenhouse gas emissions per the production unit and an increase in income, while maintaining the current input structure.

To compare the emission load of livestock products, several studies were carried out again. Initial work [61] compared load from conventional and organic cattle breeding without milk production. One kilogram of organic beef produced twice higher emissions than one kilogram of conventional meat. Another study [53] compared pork. Organic pork was again worse than conventional meat in terms of emissions. On the contrary, when comparing variants of milk, organic milk was a little emission-less burdensome than conventional milk. The latest from animal studies compared the production of eggs [62], where organic eggs produce almost 40% lower emissions than conventional eggs. Results and emission savings are summarized in the Table 2.

The higher emission load in organic farming systems is mainly due to technology of rearing and fattening when in the organic farming system, young ones are fed with breast milk while in conventional breeding, they are fed with feed. Production of breast milk causes significantly more emissions then production of crops for feed mixtures. Additionally, within conventional breeding, the emission load is divided among several products (meat, milk).

Product	Organic*	Conventional*	save BIO
milk	1,336	1,420	6%
egg	0.219	0.383	43%
beef	24,10	11,45	**-110%**
pork	6,643	5,143	**-29%**

*in kg CO_2e/kg of products(in egg study in kg CO_2e/egg)

Table 2. Emission of GHGs from the animal production (agriculture phase only)

5.1.2. Manufacturing phase

Environmentally friendly farming systems that utilize anti-erosion measures, advanced methods of nitrogen management and other measures, have the potential to sequester carbon and reduce greenhouse gas emissions [63]. This creates a positive environmental potential which may however be discarded in the following, or vice versa agricultural stage preceding, parts of the food production process which could result in a significant increase in CO_2e emissions. [64] states that within cereal production, the production of fertilizers in the pre-farming cycle makes up 35% of total emissions, while the farm stage only 27%.

Importance of pre-farming and post-farming stage can be documented by the example of potato, where [65] states the production of 0.145 kg of CO_2e in the conventional and 0.126 kg of CO_2e in organic farming system per one kilogram of potatoes. However, if we take into account also other phases (especially the processing and transport), the load resulting from potato products in relation to potatoes grows significantly. For one kilogram of peeled potatoes in the Czech Republic, it is 0.262 kg of CO_2e in conventional 0.247 kg of CO_2e in organic farming systems. However, for the manufacture of chips, it is already 2.072 kg of CO_2e in conventional and 2.271 kg of CO_2 in organic farming systems per one kilogram of finished product. And in the case of mashed potatoes, even in conventional production, it is 3.201 kg of CO_2e and in organic production 3.192 kg of CO_2e. These findings suggest that the differences between the production systems are relatively small if we compare it to the difference in CO_2e emissions between processed and unprocessed products. Another important factor is also common transport distances. Their importance is higher with the processed products that are in their life cycle more transported (besides transporting raw materials, there is still transport of semi-finished products between processing units). The transport distance is also affected by the density of processing networks and infrastructure. The results of the finished material (see Table 3) in our study [53] showed that eleven of the 22 evaluated products have better results as a conventional variety and eleven products have better result as a organic variety. This indicates a lack of potential of a manufacturing and sale network for organic products.

Group	Product	Organic*	Conventional*	save BIO
corn products	wheat	0,4593	0,4699	2%
	rye	0,3336	0,5495	39%
	wheat flour	0,6463	0,5861	**-10%**
	rye flour	0,5080	0,6737	25%
	roll	0,8100	0,7766	**-4%**
	bread	1,0431	1,0632	2%
	pasta	0,7336	0,7020	**-5%**
	rice	0,6197	0,6266	1%
vegetables products	potatoes	0,1931	0,1867	**-3%**
	peeled potatoes	0,2475	0,2624	6%
	puree	3,1918	3,2009	**0%**
	pommes	2,2714	2,0718	**-10%**
	cabbage	0,0851	0,1151	26%
	carrot	0,1158	0,1517	24%
	tomato	0,1748	0,1802	3%
	onion	0,1749	0,1285	**-36%**
	peeled onion	0,2428	0,1789	**-36%**
fruit products	apple	0,1273	0,1189	**-7%**
milk products	milk	1,4870	1,5603	5%
	yoghurt	1,7390	1,8123	4%
meat products	beef	24,5313	11,6510	**-111%**
	pork	6,7452	5,3083	**-27%**

*in kg CO_2e/kg of products

Table 3. Emission of GHGs from the final products

Besides transport distances, also the way of transportation has the influence. E.g. [63] states that significant energy savings could be achieved by rail preference which can reduce power consumption by up to half while emissions of greenhouse gases are reduced comparably. These factors, together with the production technology may, in some cases, eliminate emissions savings resulting from environmentally friendly management system. The principle of regionality which reduces unnecessary transport processes is thus superior to the principles of organic farming, since its failure may to reduce or completely eliminate the environmental potential, respectively, the emission savings resulting from organic farming,. Reducing the environmental potential can be demonstrated e.g. by the example of the production of bread in conventional and organic farming systems in the Czech Republic. Thanks to the low-volume technologies in production of bread in organic processing capacities, produced greenhouse gas emissions are much higher, so the positive effect of previous organic cultivating of wheat and flour production is eliminated [66]. Post-farming life cycle stages of agricultural products are very significant in terms of GHG emission production because within them, the emission savings generally made by organic farming in relation to conventional farming can be devalued. Assuming that the growing agricultural systems with arable land and permanent crops and grazing systems worldwide can sequester up to 200 kg C ha-1 year-1, the global carbon sequestration can reach 2.4 billion tons of CO_2e year -1. This minimum idea of conversion to organic farming would be able to lose 40% of global agricultural GHG emissions [10]. Environmentally friendly and organic farming systems are such an important tool for reducing greenhouse gas emissions.

Acknowledgements

This research study was supported by the European territorial co-operation Austria-Czech Republic 2007-2013, the Project EUS M00080– Sustainable kitchen and by the University of South Bohemia in Ceske Budejovice grant GAJU 063/2013/Z.

Author details

Zuzana Jelínková*, Jan Moudrý Jr, Jan Moudrý, Marek Kopecký and Jaroslav Bernas

*Address all correspondence to: jelinkova@zf.jcu.cz

University of South Bohemia in České Budějovice, Faculty of Agriculture, České Budějovice, Czech Republic

References

[1] Šarapatka B, Niggli U, Čížková S, Dytrtová K, Fišer B, Hluchý M, Just T, Kučera P, Kuras T, Lyth P, Potočiarová E, Salaš P. Zemědělství a krajina – cesty k vzájemnému souladu. Olomouc: Univerzita Palackého v Olomouci; 2008. 271 p. ISBN: 9788024418858

[2] Schau EM, Fet AM. LCA Studies of Food Products as Background for Environmental Product Declarations – Literature Review. International Journal of Life Cycle Assessment. 2008;13(3):255-264. DOI: 10.1065/lca2007.12.372

[3] Nemešová I, Pretel J. Skleníkový efekt a životní prostředí: Podstata, rizika, možná řešení a mezinárodní souvislosti. Praha: MŽP ČR; 1998. 70 p. SBN 80-7212-046-8

[4] Svendsen GT. How to Include Farmers in the Emission Trading System. ICROFS News. 2011;1:10-11

[5] Weinzettel J. Posuzování životního cyklu (LCA) a analýza vstupů a výstupů (IOA): vzájemné propojení při získávání nedostupných dat [thesis] Praha: ČVUT; 2008

[6] Greadel T, Allenby B. An Introduction to Life Cycle Assessment. In: Greadel T, Allenby B. (Eds.). Industrial Ecology. New Jersey: Pearson Education; 2003. p. 183-196

[7] Wagner U, Geiger B, Dreier T. Environmental Impacts and System Analysis of Biofuels. In: Proceedings of the International Conference Biomass for Energy and Industry. 8-11 june1998; Würzburg. p. 544-548

[8] Haas G, Wetterich F, Geier U. Life Cycle Assessment Framework in Agriculture on the Farm Level. International Journal of Life Cycle Assessment. 2000;5(6):345-348. DOI: 10.1065/lca2000.11.038

[9] de Backer E, Aertsens J, Vergucht S, Steurbaut W. Assessing the Ecological Soundness of Organic and Conventional Agriculture by Means of Life Cycle Assessment (LCA): A Case Study of Leek Production. British Food Journal 2009;111(10):1028-1061. DOI: 10.1108/00070700910992916

[10] Niggli U, Fliessbach A, Hepperly P, Scialabba N. Zemědělství s nízkými emisemi skleníkových plynů. Olomouc: Bioinstitut; 2011, 26 p. ISBN: 978-80-87371-11-4

[11] Baumert KA, Herzog T, Pershing J. Navigating the Numbers Greenhouse Gas Data and International Climate Policy. USA: World Resources Institute; 2005. 93 p.ISBN: 1-56973-599-9

[12] Friel S, Dangour DA, Garnett T, Lock K, Chalabi Z, Roberts I, Butler A, Butler CD, Waage J, McMichael AJ, Haines A. Public Health Benefits of Strategies to Reduce Greenhouse-Gas Emissions: Food and Agriculture. Health and Climate Change. 2009;374:2016-2025. DOI: 10.1016/S0140-6736(09)61753-0

[13] Smith P, Martino D, Cai Z, Gwary D, Janzen H, Kumar P, McCarl B, Ogle S, O'Mara F, Rice Ch, Scholes B, Sirotenko O. Agriculture. In: Metz B, Davidson OR, Bosch PR, Dave R, Meyer LA (Eds.). Climate Change 2007: Mitigation, Contribution of Working Group III to the Fourth Assessment Report of the Intergovernmental Panel on Climate Change. Cambridge: Cambridge University Press; 2007. p. 497-540

[14] Brandt US, Svendsen GT. A Project-Based System for Including Farmers in the EU ETS. Journal of Environmental Management. 2011;92(4):1121-1127. DOI: 10.1016/j.jenvman.2010.11.029

[15] Pendolovska V, Fernandez R, Mandl N, Gugele B, Ritter M. Annual European Union Greenhouse Gas Inventory 1990–2011 and Inventory Report 2013. Copenhagen: European Environment Agency; 2013, 20 p

[16] National Inventory Submission [Internet]. 2011. Available from: http://unfccc.int/national_reports/annex_i_ghg_inventories/national_inventories_submissions/items/5888.php[Accessed: 2015-02-10]

[17] Miňovský O, Krtková E, Fott P. National Greenhouse Gas Inventory of the Czech Republic. Praha: ČHMÚ; 2013. 327 p.

[18] Tukker A, Huppes G, Guinee J, Heijungs R, de Koning A, van Oers L,Suh S, Geerken T, van Holderbeke M, Jansen B, Nielsen P. Environmental Impacts of Products (EIPRO) – Analysis of the Life Cycle Environmental Impacts Related to the Final Consumption of the EU-25. Spain: Communities ; 2006. 139 p.

[19] Alluvione FB, Moretti Sacco D, Grignani C. EUE (Energy Use Efficiency) of Cropping Systems for a Sustainable Agriculture. Energy. 2011;36(7):4468-4481. DOI: 10.1016/j.energy.2011.03.075

[20] Nátr L. Země jako skleník: Proč se bát CO_2? Praha: Academia; 2006. 143 p. ISBN 80-200-1362-8

[21] Moudrý (jr.) J, Konvalina P, Moudrý J, Kalinová, J. Základní principy ekologického zemědělství. České Budějovice: Jihočeská univerzita v Českých Budějovicích; 2007, 40 p.

[22] LaSalle T, Hepperly P. Regenerative Organic Farming: a Solution to Global Warming. Kutztown: Rodale Institut; 13 p.

[23] Mondelaers K, Aertsens J, van Huylenbroeck G. A Meta-Analysis of the Differences in Environmental Impacts between Organic and Conventional Farming. British Food Journal. 2009;111(10):1098-1119. DOI: 10.1108/0007070091099292

[24] Dalgaard T, Ferrari S, Rambonilaza M. Introduction: Features of Environmental Sustainability in Agriculture: Some Conceptual and Operational Issues. International Journal of Agricultural Resources, Governance and Ecology. 2006;5(2/3):107-115. DOI: 10.1504/IJARGE.2006.009177

[25] Wallgren C, Höjer M. Eating Energy – Identifying Possibilities for Reduced Energy Use in the Future Food Supply System. Energy Policy. 2009;37(12):5803–5813.DOI: 10.1016/j.enpol.2009.08.046

[26] Lal R. Soil Carbon Sequestration Impacts on Global Climate Change and Food Security. Science. 2004;304:1623–1627. DOI: 10.1126/science.1097396

[27] Monni S, Syri S, Pipatti R, Savolainen I. Extension of EU Emissions Trading Scheme to Other Sectors and Gases: Consequences for Uncertainty of Total Tradable

Amount. Water Air & Soil Pollution: Focus 2007;7(4-5):529-538. DOI: 10.1007/978-1-4020-5930-8_9

[28] Finnveden G, Hauschild MZ, Ekvall T, Guninée J, Heijungs R., Hellweg S, Koehler A, Pennington D, Suh S. Review. Recent Developments in Life Cycle Assessment. Journal of Environmental Management. 2009;91(1):1-21. DOI: 10.1016/j.jenvman. 2009.06.018

[29] Thomassen MA, de Boer IJM. Evaluation of Indicators to Assess the Environmental Impact of Dairy Production Systems. Agriculture. Ecosystems and Environment. 2005;111: 185-199. DOI: 10.1016/j.agee.2005.06.013

[30] Remtová K. Posuzování životního cyklu – Metoda LCA. Praha: MŽP; 15 p.

[31] Jørgensen A, Le Bocq A, Hauschild MZ. Methodologies for Social Life Cycle Assessment – a Review. The International Journal of Life Cycle Assessment. 2008;13(2): 96-103

[32] Kočí V. Posuzování životního cyklu – Life cycle assessment LCA. Chrudim: Ekomonitor; 2009. 263 p. ISBN 978-80-86832-42-5

[33] Garnett T. (2003): Wise Moves: Exploring the Relationship Between Food, Transport and CO_2. London: Transport 2000 Trust; 2003. 112 p

[34] Thomassen MA, van Calker KJ, Smiths MCJ, Iepema GL, de Boer IJM. Life Cycle Assessment of Conventional and Organic Milk Production in the Netherlands. Agricultural Systems. 2008;96(1-3): 95–107. DOI: 10.1016/j.agsy.2007.06.001

[35] Goedkoop M, Oele M, Leijting J., Ponsioen T., Meijer E. Introduction to LCA with SimaPro. Amersfoort: PRé; 2013. 80 p.

[36] ČSN EN ISO 14 044. Environmentální management – Posuzování životního cyklu – Požadavky a směrnice. Praha: Český normalizační institut; 2006.

[37] Basset-Mens C, van der Werf HMG. Scenario-based Environmental Assessment of Farming Systems: the Case of Pig Production in France. Agriculture, Ecosystems & Environment. 2005;105(1-2):127-144. DOI: 10.1016/j.agee.2004.05.007

[38] Brentrup F. Life Cycle Assessment to Evaluate the Environmental Impact of Arable Crop Production [thesis]. Hannover: University of Hannover;2003

[39] Roy P, Nei D, Orikasa T, Xu O, Okadome H, Nakamura N, Shiina T. A Review of Life Cycle Assessment (LCA) on Some Food Products. Journal of Food Engineering. 2009;90(1):1-10. DOI: 10.1016/j.jfoodeng.2008.06.16

[40] De Koeijer TJ, Wossink GAA, Struik PC, Renkema JA. Measuring Agricultural Sustainability in Terms of Efficiency: The Case of Dutch Sugar Beet Growers. Journal of Environmental Management. 2002;66(1):9-17. DOI: 10.1006/jema.2002.0578

[41] Berlin D, Uhlin HE. Opportunity Cost Principles for Life Cycle Assessment: Toward Strategic Decision Making in Agriculture. Progress in Industrial Ecology: An International Journal. 2004;1(1-3):187-202. DOI: 10.1504/PIE.2004.004678

[42] Casey JW, Holden NM. Greenhouse Gas Emissions from Conventional, Agri-Environmental Scheme, and Organic Irish Suckler-Beef. Journal of Environmental Quality. 2006;35:231-239. DOI: 10.2134/jeq2005.0121

[43] Williams AG, Pell E, Webb J, Tribe E, Evans D, Moorhouse E, Watkiss P. Comparative Life Cycle Assessment of Food Commodities Procured for UK Consumption through a Diversity of Supply Chains. Cranfield: Cranfield University, 2008; 78 p.

[44] Jungbluth N, Tietje O, Scholz RW. Food Purchases: Impacts from the Consumers' Point of View Investigated with a Modular LCA. International Journal of Life Cycle Assessment. 2000;5(3):134-142. DOI: 10.1007/BF02994077

[45] Nemecek T, Erzinger S. Modelling Representative Life Cycle Inventories for Swiss Arable Crops. International Journal of Life Cycle Assessment, 2005;10(1):68-76. DOI: 10.1065/lca2004.09.181.8

[46] Kočí V. Na LCA založené srovnání environmentálních dopadů obnovitelných zdrojů energie – Odhad LCA profilů výroby elektrické energie z obnovitelných zdrojů energie v ČR pro projekt OZE – RESTEP. Praha: VŠCH; 2012. 111 p.

[47] Rebitzer G, Ekvall T, Frischknecht R, Hunkeler D, Norris G, Rydberg T, Schmidt WP, Suh S, Weidema BP, Pennington DW. Life Cycle Assessment Part 1: Framework, Goal and Scope Definition, Inventory Analysis, and Applications. Environment International. 2004;30(5):701-720. DOI: 10.1016/j.envint.2003.11.005

[48] Owens JW. LCA Impact Assessment Categories. International Journal of Life Cycle Assessment. 1996;1(3):151-158. DOI: 10.1007/BF02978944

[49] Heiskanen E. Managers' Interpretations of LCA: enlightenment and Responsibility or Confusion and Denial? Business Strategy and the Environment. 2000;9(4):239-254. DOI: 10.1002/1099-0836200007

[50] De Klein C, Novoa SAR, Ogle S, Smith KA, Rochette P, Wirth TC. N_2O emissions from managed soil and CO_2 from lime and urea application. In: Eggleston S, Buendia L. Miwa K, Ngara T, Tanabe K. 2006 IPCC Guidelines for Nation Greenhouse Gas Inventories. Volume 4 Agriculture, Forestry and Other Land Use. Geneva: IPCC; 2006. P. 1-54.

[51] Dong H, Mangina J, McAllister TA,Hatfield JL, Johnson DE, Lassey KR, Aparecida de Lima M, Romanovskaya A. Emissions for Livestock and Manure Management. In: Eggleston S, Buendia L. Miwa K, Ngara T, Tanabe K. 2006 IPCC Guidelines for Nation Greenhouse Gas Inventories. Volume 4 Agriculture, Forestry and Other Land Use. Geneva: IPCC; 2006. P. 1-37.

[52] Moudrý (jr.) J, Jelínková Z, Moudrý J, Bernas J, Kopecký M, Konvalina P. Influence of farming systems on production of greenhouse gases emissions within cultivation of selected crops. Journal of Food, Agriculture & Environment. 2013;11(3-4):1015-1018.

[53] Jelínková Z, Moudrý (jr.) J, Moudrý J, Bernas J, Kopecký M, Konvalina P. LCA method – tool for food production evaluation. Lucrari stiintifice: seria agronomie. 2013;57(1):23-27.

[54] Moudrý (jr.) J, Jelínková Z, Moudrý J, Kopecký M, Bernas J. Production of greenhouse gases within cultivation of garlic in conventional and organic farming system. Lucrari stiintifice: seria agronomie. 2013;56(2):15-19.

[55] Williams AG, Audsley E, Sandars DL. Determining the Environmental Burdens and Resource Use in the Production of Agricultural and Horticultural Commodities. Cranfield: Cranfield University; 2006. 97p.

[56] Dyer JA, Desjardins RL. The Impact of Farm Machinery Management on the Greenhouse Gas Emissions from Canadian Agriculture. Journal of Sustainable Agriculture. 2003;22(3):59-74. DOI: 10.1300/J064v22n03_07

[57] Smith P, Martino D, Cai Z, Gwary D, Janzen H, Kumar P. Greenhouse Gas Mitigation in Agriculture. Biological Sciences. 2008;363(1492):789-813. DOI: 10.1098/rstb. 2007.2184

[58] Stolze M, Piorr A, Haring A, Dabbert S. The Environmental Impacts of Organic Farming in Europe. Stuttgart: University of Hohenheim; 2000. 127 p.

[59] Paustian K, Babcock B, Hatfield J, Lal R, McCarl B, McLauglin S, Mosier A, Rice C, Robertson GP, Rosenberg NJ, Rosenzweig C, Schlesinger WH et al. Agricultural Mitigation of Greenhouse Gases: Science and Policy Options. Armes: CAST; 2004. 120 p.

[60] Cederberg C, Mattsson B. Life Cycle Assessment of Milk Production – a Comparison of Conventional and Organic Farming. Journal of Cleaner Production. 2000:8(1): 49-60.

[61] Plch R, Jiroušková Z, Cudlín P., Moudrý J. The comparison of conventional beef production and bio-production using the method of life cycle assessment. In: 3rd scientific conference theme: New findings in organic farming research and their possible use for Central and Eastern Europe: November 2011; Olomouc: Bioinstitut; 2011. p. 90-94

[62] Moudrý J, Jelínková Z, Jarešová M, Plch R, Moudrý J, Konvalina P. Assessing greenhouse gas emissions from potato production and processing in the Czech Republic. Outlook on Agriculture. 2013;42(3):179-183. DOI: 10.5367/oa.2013.

[63] West TO, Marland G. A Synthesis of Carbon Sequestration, Carbon Emissions and Net Carbon Flux in Agriculture: Comparing Tillage Practices in the United States. Agriculture, Ecosystems and Environment. 2002;91(1-3):217-232. Doi:10.1016/S0167-8809(01)00233-X

[64] Biswas WK, Barton L, Carter D. Global Warming Potential of Wheat Production in South Western Australia: a Life Cycle Assessment. Water and Environment Journal. 2008;22(3): 206-216.DOI: 10.1111/j.1747-6593.2008.00127.x

[65] Moudrý J, Jelínková Z, Jarešová M, Plch R, Moudrý J, Konvalina P. Assessing greenhouse gas emissions from potato production and processing in the Czech Republic. Outlook on Agriculture. 2013; 42(3): 179-183. DOI: 10.5367/oa.2013.0138

[66] Moudrý (jr.) J, Jelínková Z, Plch R, Moudrý J, Konvalina P, Hypšler R. The emissions of greenhouse gases produced during growing and processing of wheat products in the Czech Republic. Food, Agriculture and Environment. 2013; 11(1): 1133-1136. ISSN:1459-0255

8

The Choice between Economic Policies to Face Greenhouse Consequences

Donatella Porrini

Additional information is available at the end of the chapter

Abstract

In the past few years, unstable and extreme weather patterns are increasingly occurring as phenomena of climate change, and the link to greenhouse gas (GHG) emissions is scientifically accepted. From an economic point of view, extreme weather patterns are causing major damage to health, property, and business.

In this chapter, following an economic analysis of law (EAL) approach, the issue of the comparison between the alternative environmental economic policies is analyzed starting from the consideration that the emissions of GHGs originate market failures: the environment appears as a "public good" that may not be appropriated and has no market price; the damage to the environment is a case of "externality," where it is fully or partly a social cost that is not internalized into the accounts of the parties causing it. In the EAL literature, an environmental policy instrument has been seen as it may play a role in correcting malfunction and subsequent inefficiencies.

In the first part of the chapter, we intend to revise the traditional analysis of the choice of environmental policies. The following part deals with the comparison between tax and tradable permit systems. Then the role that can be played by the insurance sector is considered. The different policy instruments are considered in the framework of climate as an economic global public good. And, finally, some conclusive remarks are presented in relation to the COP 21 conference in Paris in terms of the future policies against GHG effects.

Keywords: climate change, environmental policy choice, insurance, greenhouse gasses, COP 21

1. Introduction

Over the last centuries, climate change has been raised as a very important issue all over the world. The change in climate results from an increase in the earth's average atmospheric temperature, which is usually referred to as global warming. It may be due to both natural and human causes, especially greenhouse gas (GHG) emissions.

In response to increasing scientific evidence that human activities are contributing significantly to global climate change [1], decision makers are devoting considerable attention to public policies to reduce GHG emissions and thereby prevent or reduce such change.

The policies span a range of regulatory approaches. The main alternative is between command and control (CAC) and market-based (MB) instruments, and a relevant role can also be played by the insurance sector.

This chapter aims to describe the traditional theory on the choice of environmental policies following an economic analysis of law (EAL) approach (Section 2) to analyze the comparison between tax and tradable permit systems (Section 3), to outline the role of the insurance sector (Section 4), and to consider the different policy instruments in a context of economic global public goods. The final objective is to take into account the future COP 21 conference in Paris in terms of the choice of policy instruments against GHG effects.

2. The traditional issue of the choice of environmental policies

The problem of the choice of environmental policy instruments has been an issue since Pigou [2] analyzed the need for state intervention when private costs diverge from social costs and suggested that the solution would be to internalize the externalities through taxation.[1] Coase [4] criticized the proposed state intervention and affirmed that there is no reason to suppose that governmental regulation is called for simply because the problem is not very well handled by the market or the firm.

The ensuing debate has been conducted along these two opposite views: on the one hand, the supporters of the idea that the choice of policy instruments is to be applied as a public matter and the state, as policy designer, should select the optimal instruments and take responsibility for their imposition in the public interest, but, on the other, the supporters of MB instruments are trying to fight a battle against a sort of "anti-market" mentality based on a reluctance to apply MB instruments [5].

So the problem would be to compare the instruments that can be considered public oriented and those that can be considered market oriented, where the former is characterized by a public agency that defines a conduct rule and provides an enforcement system and the latter is characterized by instruments based on market mechanisms stimulating the conduct of the firm indirectly and by a private administration and enforcement system.

---

1 See the documents of the "Pigou club" [3].

Following an EAL approach, traditionally regulatory systems originate from the presence of market failure: in our specific case, the environment appears as a "public good" that may not be appropriated and has no market price; the damage to the environment is a case of "externality," in that it is fully or partly a social cost that is not internalized into the accounts of the parties causing it.[2] So the comparison of different instruments can consider how they may play a role in correcting malfunction and subsequent inefficiencies [7].

In this way, we can move from the theoretical definition of the efficiency of different instruments to their practical, and so direct, potential to achieve concrete objectives. In particular, the objectives that emerge as relevant in judging the practical efficiency of environmental policies are the following: first is the prevention in the sense of providing incentives for firms to improve safety standards, and second is the connection with technological change in the sense of encouraging firms to adopt lower risk technologies.

The first kind of environmental instrument is the so-called CAC that is characterized by a public agency that provides a definition of conduct rules and enforcement system. Thus, they could be defined as public-oriented instruments, which require the use of a particular technology or the observation of a performance standard, authorizing for the maximum amount that a source can emit.

CAC is divided into two phases as follows: "command", which sets a standard based on the maximum level of permissible emissions, and "control", which monitors and enforces the standard.

Regarding standards, they can be classified into two: ambient standards, which fix a minimum desired level of air or water quality or a maximum level of emissions that must be maintained, and emission standards, which fix a maximum level of permitted emissions that can be performance based, setting emission limits that each firm is allowed, or technology based, specifying the best technology to be used.[3]

As to the US experience with this kind of regulation, the activity of the Environmental Protection Agency (EPA) provides a clear example of regulation by an independent environmental authority. This agency performs its tasks through the setting of preventive standards, the enforcement of polluting emission thresholds, and the performance of inspections and, possibly, of actions brought to the federal courts. We cannot mention the European Community experience given that a standard setting system has not been established at a European level (but at national level) and that the European Environmental Agency (EEA) has only a very limited role.[4]

---

2 Economists consider environmental policies within the framework of the category of externalities, as evidenced by Cropper and Oates [6]: "The source of basic economic principles of environmental policy is to be found in the theory of externality."

3 As Cropper and Oates [6] explain"…The determination of environmental policy is taken to be a two-step process: first, standards or targets for environmental quality are set, and, second, a regulatory system is designed and put in place to achieve these standards. This is often the environmental decision making proceeds. Under the Clean Air Act, for example, the first task of the EPA was to set standards in the form of maximum permissible concentration of the major air pollutants. The next step was to design a regulatory plan to attain these standards air quality."

The choice to develop a CAC regulatory system is based on the advantage of centralized agencies to assure a cost-effectiveness calculation on the base of the expected damage and of the marginal cost of different level of preventive care. The centralized structure presents the advantage to provide a continual oversight of problems and a broad array of regulatory tools.

Following the traditional EAL approach, well-defined standards generate the correct incentive for the firm to act with caution and make the best production and prevention decisions [8, 9].

CAC instruments use to be compared with MB instruments that are characterized by a private administration and enforcement system and stimulate indirectly the behavior of the firm. There are essentially two different types of those instruments: taxes that are fees imposed on emitters proportionate to the total amount of emissions released into the environment (they could be divided into emission charges, product charges, and user charges) and marketable (or tradable) permit systems that provide a fixed number of permits equal to the allowed total emissions, distributing them among polluting firms in a specific area.

The two types of MBI instruments can be seen in the following two different approaches: on one side, taxes follow a price approach because producers adjust the quantity, given a fixed price put on emissions; on the other side, a tradable permit system follows a quantity approach because the price is adjusted according to supply and demand, given a maximum quantity of emissions allowed.

3. Carbon tax versus tradable permit system

A carbon tax is a particular tax based on GHG emissions generated by burning fuels and biofuels, such as coal, oil, and natural gas. It has been introduced with the main goal to level the gap between carbon intensive (firms based on fossil fuels) and low carbon intensive (firms that adopt renewable energies) sectors. Due to the introduction of that form of taxation, the relative prices of goods and services will change; the emissions of intensive goods will be more expensive, whereas the emissions of less intensive goods will be lower. Thus, carbon tax provides a strong incentive for individuals and firms to adjust their conduct, resulting in a reduction of the emissions themselves. Hence, by reducing fuel emissions and adopting new technologies, both consumers and businesses can reduce the entire amount they pay in carbon tax.

A tradable permit system is defined as quantity-based environmental policy instrument. The regulatory authority stipulates the allowable total amount of emissions (cap) and the right to emit becomes a tradable commodity. Under a cap-and-trade system, prices are allowed to fluctuate according to market forces. Thus, the price of emissions is established indirectly. Permits could be allocated to firms through auction or free allocation.

---

4 The EEA was set up as a legally independent community body under council regulation (EEC) 1210/90. The EEA's core task is to provide decisionmakers with the information needed for making sound and effective policies to protect the environment and support sustainable development.

Similarly to other environmental taxes, carbon taxes are defined as priced-based policy instruments for the correlated effects to increase the price of certain goods and services, thereby decreasing the quantity demanded. On the other side, tradable permits are defined as quantity-based environmental policy instrument. Although both policy instruments are MB, their implementation is different: carbon taxes fix the marginal cost for carbon emissions and allow quantities emitted to adjust, whereas tradable permits fix the total amount of carbon emitted and allow price levels to change according to market forces.

Which is better? There is no simple yes or no answer, and the policies are not necessarily mutually exclusive. Several important advantages and drawbacks of the respective policies are outlined later.

A well-functioning emission trading system allows emission reductions to take place wherever abatement costs are lowest, regardless of international borders. As costs associated with climate change have no correlation with the origin of carbon emissions, the rationale for this policy approach is that an emission trading system allows to fix a certain environmental outcome and the companies are called to pay a market price for the rights to pollute. This is the reason why an emission trading system is suitable for international environmental agreements, such as the Kyoto Protocol, specifically for the characteristic that a defined emission reduction level can be easily agreed between states.

Emission trading is more appealing to private industry because, by decreasing emissions, firms can actually profit by selling their excess GHG allowances. Creating such a market for pollution could potentially drive emission reductions below targets.

A carbon tax would offer a broader scope for emission reductions [10]. A system of tradable permits entails significant transaction costs, which include search costs, such as fees paid to brokers or exchange institutions to find trading partners; negotiating costs; approval costs; and insurance costs. Conversely, taxes involve little transaction cost over all stages of their lifetime.

Carbon taxes are dynamic economic instruments that offer a continuum incentive to reduce emissions. In fact, technological and procedural improvements and their subsequent efficient diffusion lead to reductions in tax payment. In addition, trading systems are able to self-adjust because emission goals will be easier to meet; there will be a decrease in permits' demand and in their price but not as rapidly as taxes.

The implementation of an emission trading system is very complicated and requires technical steps, including treatment of sinks, monitoring, and enforcement. On the other hand, taxation is a very well-known instrument by policy makers, not very costly because it does not require monitoring and enforcement organization.

The revenue from carbon taxes can be used into the economy to reduce income taxes or levies on labor or capital investment. This may be part of a national or international reform of the taxation systems with the effects to shift the tax burden from "goods" like labor to "bads" like pollution.

In Table 1, we have summarized the main differences among CAC, carbon tax, and tradable permit system.

	Command and control	Carbon tax	Tradable permit system
Certainty over CO_2 price or cost?	No	Yes. The tax establishes a well-defined price	No. But price volatility can be limited by design features, such as a safety valve (price cap) or borrowing
Certainty over emissions?	No. Regulating the rate of emissions, the level uncertain leaves	No. Emissions vary with prevailing energy demand and fuel prices	Yes, in its traditional form. No, with the use of additional cost containment mechanisms
Efficiently encourages least-cost emission reductions?	No, but tradable standard is more efficient than non-tradable standard	Yes	Yes
Ability to raise revenue?	No	Yes. Results in maximum revenue generation compared with other options	Traditionally no
Incentives for R&D in clean technologies?	Yes and no. Standards encourage specific technologies but not Brod innovation	Yes. Stable CO_2 price is needed to induce innovation	Yes. However, uncertainty over permits' prices could weaken innovation incentives
Harm to competiveness?	Somewhat. Regulations increase the cost of manufacturing, but, unlike taxes of tradable permits, do not raise the price of fossil energy	Yes, though if other taxes are reduced through revenue recycling, competitiveness of the broader economy can be improved	Yes (as with a tax), but giving firms free allowances offsets potentially harmful effects on profitability
Practical or political obstacles to implementation?	Yes. Setting the level of standards is difficult	Yes. New taxes have been very unpopular	Yes. Identifying a reasonable allocation and target is difficult
New institutional requirements?	Minimal (unless tradable)	Minimal	Yes, but experience with existing trading programs suggests that markets arise quickly and relatively inexpensively

Source: Parry and Pizer [11].

Table 1. Command and control versus carbon tax versus tradable permit system

4. Insurance as an environmental policy instrument

A relevant role could be played by the insurance sector in the choice of political economic instruments for climate change.

As mentioned in Section 1, following the EAL approach, the emitters of GHGs externalize the true costs of their contribution to climate change, and this implies the need to recover these costs, which manifest through both the costs of impacts and the costs to prevent impacts. Because the insurance sector is the world's largest industry, the response of insurers to the broader climate change challenge will no doubt be the key to, at least partially, solve this internalization problem.

Generally, insurers can indirectly stimulate prevention behavior in their customers. In the case of climate change, they can play two primary roles through their insurance products. The first role is played in supplying and pricing traditional insurance coverage for damage deriving from climate change by the promotion of actions of businesses and individuals to align policy holders with climate-positive behaviors. The second is played in providing capital to new ventures and reducing the financial risk to investors by facilitate the creation of new markets and services that will help to solve the climate change problems.

In both the roles, insurance companies deal with the management of risk activities. In fact, climate change risk is managed by the insurers through the risks they accept from clients, given that climate change causes an increase in intensity and spreading the distribution of extreme weather events with resultant effects on property claims that could be catastrophic.[5]

Traditionally, insurance is the main mechanisms available to individuals and business to manage the financial consequences of risky events, such as natural hazards like windstorms and floods. Insurance companies used to work making each individual or business pay a premium to protect themselves against an uncertain loss. The premium is calculated by pooling risks across a large and diverse population considering the pool's expected losses.

In this way, the insurance industry provides a coverage for climate change consequences because climate experts predict changes in the intensity and the distribution of extreme weather events and of the resulting risk of catastrophic property claims.

But specifically, for what concerns the consequences of climate change, many problems arise to provide insurance coverage [13].

First, climate change is connected to global weather patterns that may increase the potential for losses so large. As more severe weather becomes more common and overall variability of conditions increases, there is a threat for the solvency of insurance companies.

Second, because uncertainties in assessing climate change's impacts are high and affect property, casualty, business interruption, and health; as a result, risk has significant ambiguous components and insurers are both likely to charge a significantly higher premium or avoid insuring the risk entirely.

Third, climate change–related risks may be correlated and create a skewed risk pool that could increase the probability of extremely large losses and also risks not well-distributed across existing insurers.

---

5 Insurance coverage can be also connected with the liability system. But, for what concerns the consequences of climate change, such as impacts of property damage from extreme weather events, the assignment of liability is a very complex topic. See Reference [12].

Finally, as a result of insurers' uncertainty aversion and need to protect against extremely large losses from single or related events, it is not clear that insurers will be willing to insure against some climate change–related risks at a price that policy holders are willing to pay.

Climate change will affect, and in some cases is already affecting, most major types of insurance products. Insurers will feel the impact of climate change on property and casualty insurance, where the insurer bears the risk of a loss suffered directly by the policy holder. These property and casualty claims include not only damage to insured property as a direct result of weather but also claims for business interruptions and other consequences of weather-induced events. Also, health and life insurers are going to face increasing costs.

Sometimes, insurance companies are involved in the systems of compensation fund that have been established in some countries. This is the case of special government disaster funds within the target to promote framework of contingency measures to cover the economic costs of natural disasters [14].

Other insurance products are the so-called "financial responsibility" products. This term includes the tools that require companies to demonstrate to have sufficient financial resources for eventual future environmental damage that may arise through their activities. In its common implementation, financial responsibility implies that a production activity is authorized only if companies can prove that they will be able to financially cover economic claims, for example, using surety bonds, cash accounts, deposit certificates, self-insurances, and corporate guarantees.

In the past few years, the insurance industry has developed financial products suitable for dealing with climate change-related risks in the direction to play a role far beyond simply compensating climate change's victims for their losses ex post. The activity of the insurance has become relevant as a political economic instruments within an ex ante strategy to financially manage large-scale catastrophes, as a complement of ex post instruments for the compensation of disaster losses.

Insurance industry is also developing alternative risk transfer products, given that conventional reinsurance arrangements may in future cover a smaller proportion of total losses, and there may be insufficient capital available to insurance markets to cover these losses [15].

A first kind of these insurance products are called "catastrophe bonds" and consist in securitizing environmental risks in bonds, which could be sold to high-yield investors. The catastrophe bonds are able to transfer risk to investors that receive coupons that are normally a reference rate plus an appropriate risk premium. By these products, insurance companies limit risk exposure transferring natural catastrophe risk into the capital markets. In this way, with the involvement of the financial markets, their global size offers enormous potential for insurers to diversify risks.

Weather derivatives are another kind of financial instrument used by companies to hedge against the risk of weather-related losses. Weather derivatives pay out on a specified trigger, for example, temperature over a specified period rather than proof of loss. The investor providing a weather derivative charges the buyer a premium for access to capital. If nothing happens, then the investor makes a profit.

With this kind of financial products, the insurance industry tries to reach two goals. First, there is the need for extra capital and to spread risks beyond the insurance sector. Particularly, cat bonds are used to spread insurance risk in the financial sector. The second goal is to improve the accuracy and the resolution of hazard data and the likely impacts on climate change with the involvement of financial market forecast ability.

The insurance industry can act to tackle the consequences of climate change by playing its part in climate change mitigation, through the promotion of ways to reduce GHG emissions. Insurers are also well placed to help society to adapt to the impacts of climate change, by promoting the effective limitation and management of risks from extreme weather-related hazards.

Individuals and companies that buy insurance products could be stimulated to address climate change-seeking mechanism to facilitate mitigation of GHG emissions and adaptation to the inevitable impacts of climate change [16]. In addition, the insurance companies themselves are motivated to take significant actions to mitigate GHG emissions and increase adaptive capacity to reduce overall uncertainty and other barriers to insurability and are also motivated to limit the insurers' potential exposure to catastrophic risks in excess of their capacity to avoid the potential for property and liability claims in excess.

The insurance industry is also developing new products that would have the consequences to stimulate the adaptation to climate change. On the one hand, products help to create the conditions for active adaptation to build physically resilient communities; on the other hand, products provide capital and liquidity to help communities to cope with losses caused by climate change catastrophes.

Insurance products can be designed in a way for which simply their price reflects the level of climate change-related risk assumed by the insurer. These products, rewarding behavior that reduces risk of financial losses, encourage adaptive behavior. For example, insurance products that incorporate these features provide a premium discounts on property insurance for climate-resilient commercial or residential buildings.

The recent tendency to supply coverage with differential premiums to customers depending on their level of protection from losses caused by weather-related disasters can be seen as a clear opportunity for insurers to reduce their own overall and maximum possible loss exposure while promoting communities overall resilience in the face of climate change's impacts. So, more often, policies include discounts for businesses or homeowners that have taken specific steps to ensure their buildings resistant to floods or other hazards. Moreover, insurance companies also condition their policies on compliance with laws such as building codes, thus playing a role in enforcing laws that promote climate change resilience.

The second kind of insurance products that stimulate adaptation is based on the availability of capital to cope with catastrophes. These are financial arrangements intended to bring needed capital that will reduce the risk that could derive from future climate-related hazards for those who are most likely to be in peril. These products can be defined as adaptation oriented because they help to build the capacity of nations, communities, and businesses to cope with climate change impacts.

Figure 1 shows the different roles that can be played by insurance sector as a form of environmental policy instruments.

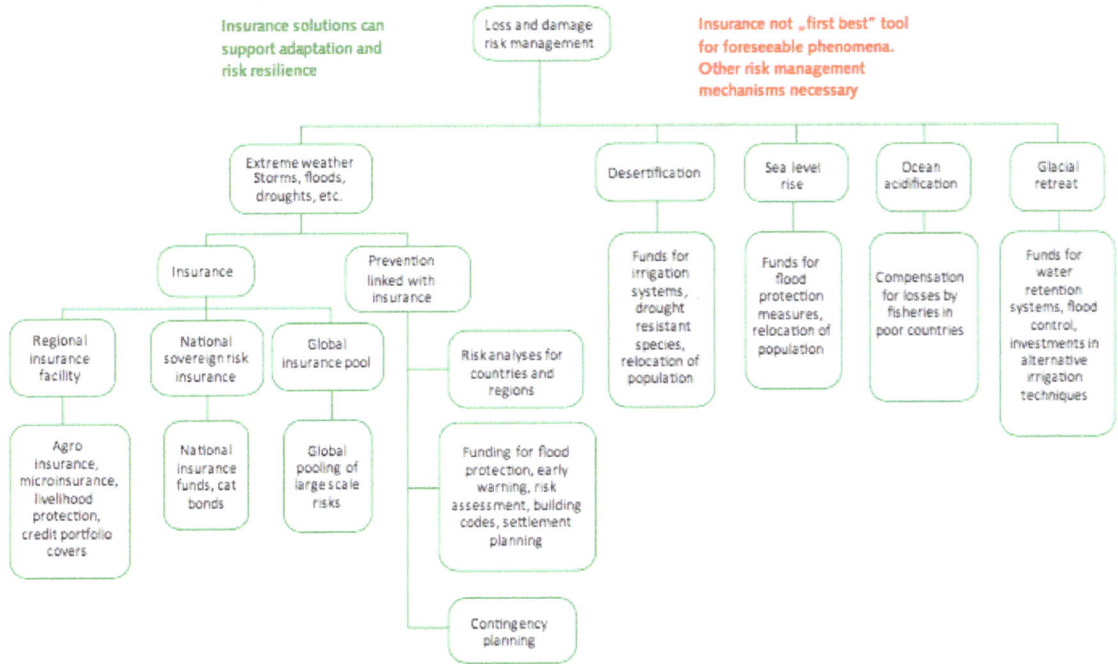

Source: MCII [17].

Figure 1. The different roles of the insurance sector in the framework of climate change consequences

5. Policies choice in the case of climate change and the concept of "economic global public goods"

Dealing with the consequences of GHG emissions in terms of climate change implies a definition of "economic global public goods" that can be defined as goods with economic benefits that extend to all countries, people, and generations [18]. Following the EAL approach, they are special case of externalities with a global dimension.

Climate is clearly "global" in both causes and consequences; moreover, the emissions of GHG have effects on global warming independently of their location, and local climatic changes are completely linked with the world climate system.

In addition, global warming is characterized by other important features that imply some difficulties in the implementation of the instruments provided by the standard economic theory of policy choice. First, we cannot determine with certainty both the dimension and the timing of climate change and the costs of the abatement of emissions. Second, the effects of GHG concentration in the atmosphere on climate are intergenerational and persistent across time. Finally, it emerges a relevant equity issue among countries because industrialized countries have produced the majority of GHG emissions, but the effects of global warming will be much more severe on developing countries. In other words, the countries that have more responsibilities will face less consequence in the future and vice versa.

There are major governance issues involved in dealing with global public goods because global coordination is required [19]. With economic public goods, it is difficult to determine and reach agreement on efficient policies because they involve estimating and balancing costs and benefits where neither is easy to measure and both involve major distributional concerns. When dealing with economic public goods like global warming, it is necessary to reach through governments to the multitude of firms and consumers who make the vast number of decisions that affect the ultimate outcomes.

Because global warming is a global public good, the key environmental issue is global emissions, and the key economic issue is how to balance costs and benefits of global emission reductions. Climate change depends only on total GHG emissions and the time path of emissions not on the geographic location of emissions. Moreover, the impacts depend primarily on cumulative emissions that remain in the atmosphere not on the annual flow of emissions.

It is a global issue to decide what the distribution of emission reductions among countries should be and how the costs should be allocated, together with the need for differences among high- and low-income countries, high- and low-emitting countries, and high- and low-vulnerability countries.

Given the global nature of climate change, it is easy to understand the necessity of an action at international level, in order to efficiently implement the different policy instruments that we have analyzed earlier in this chapter.

First, the instruments based on tax mechanism need a method of coordinating policies among countries. In the international environment, it could assume the form of either an international tax or a harmonized domestic tax system. In the case of an international tax, the nations (and not the firms) pay the tax to an international agency, which receives and redistributes the tax revenues. On the other hand, in the case of harmonized domestic tax, the international community should negotiate an agreed level of a domestic emission tax, establishing adequate compensation for the losing countries from the gaining countries.[6]

Second, it is possible to establish an agreement that sets quantitative limits of emissions and allocates emission permits to firms (or States) but allows to trade among countries, in order to minimize abatement costs. The starting allocation of permits can be set through either an auction or a grandfather allocation. Under an auction, government (or the international community) sells the emission permits, whereas under the grandfather rule, the allocation of emission permits is based on historical records.

---

6 Nordhaus [20] hypothesizes the institution of an harmonized carbon tax (HCM), essentially equivalent to a "dynamic Pigouvian pollution tax for a global public good" and points out 10 different reasons to prefer it to a quantitative approach.

In the global-warming context, quantitative limits set targets on the time path of GHG emissions of different countries. Countries then can administer these limits in their own fashion, and the mechanism may allow transfer of emission allowances among countries, as is the case under the Kyoto Protocol.

The European Union Emissions Trading Scheme (EU-ETS) – the world's most extensive carbon pricing market – has now been in operation for 10 years. The EU-ETS was developed as a way of meeting the EU's GHG emission reduction targets in the most efficient and cost-effective manner. To do so, the EU-ETS sets a limit (a cap) on the total emissions; certain EU sectors (mostly heavy industry and aviation) are allowed to use during predefined trading periods. Permits are then distributed among polluters where one permit equals one ton of carbon dioxide equivalent. These permits can then be traded between market participants. As such, the total amount of pollution is set by an external authority, but market participants determine the permit allocation, thereby optimizing efficiency.

In addition, the involvement of the insurance sector as an efficient policy instruments needs a global approach.

Insurance sector can contribute to develop risk management strategy to minimize climate change consequences on an urgent basis to prevent further escalation of global warming. However, insurance needs to be a part of the overall policy of mitigation and adaptation that aims at reducing the severity of many impacts that could result from climate change if current adverse conditions prevail.

In order to organize their own operations to the new challenge, the insurance industry should include climate change risk in its internal governance procedures, in line with the existing financial corporate risk identification. To enable insurance companies to play a responsible role in tackling climate change consequences, they require a reliable, transparent, and inter-national coordinated policy framework as well as long-term, appropriated GHG emission reduction goals coordinated at an international level.

6. Final remarks: COP 21 and the issue of the linkage of different national policies

In the COP 21 meeting in Paris at the end of the 2015, global climate policy has faced the tension between the efficiency benefits of uniform global policy and national and regional variation in tastes for differing policies. Although climate negotiations, going back to the framework convention, have had a coordinated global policy as their goal, it could be that we will head toward a less coordinated system of local, national, or regional policies.

In reality, as we have analyzed in the chapter, different countries are undertaking different policies ranging from CAC to MB approaches, such as carbon tax and tradable permit systems.

Variations in policies, although catering to local tastes and preferences, can lead to substantial inefficiencies, and the target will be to reach an optimal degree of policy homogenization.

There is a large literature on the importance of linking economic policies to face GHG consequences [21].

Some authors argue that the basic approach underlying emission reduction credit systems like the Kyoto Clean Development Mechanism (CDM) can be extended to create linkage opportunities in diverse emission control systems in ways that do not necessarily suffer from the shortfalls of the current CDM. Moreover, while emission reduction credit systems are designed to work with MB systems like tradable permits, they describe ways in which it can interact with tax systems as well as certain regulatory systems [22].

By clarifying the opportunities and challenges of insurance as an instrument for adaptation and outlining a practical way forward, it is hoped that this discussion contributes to the opportunities in adopting a comprehensive adaptation strategy that enables risk management and insurance through the funding of a global adaptation strategy [17].

In the near future, the challenge for COP 21 to reduce GHG emissions is to try to link heterogeneous climate policy instruments as a way to reach the solution of climate change issue in the long term.

Author details

Donatella Porrini*

Address all correspondence to: donatella.porrini@unisalento.it

Universita` del Salento, Lecce, Italia

References

[1] IPCC. Guidelines for national greenhouse gas inventories. Available from www.ipcc-nggip.iges.or.jp/public/2006gl [Accessed: 2015-10-10].

[2] Coase R. The problem of social cost. Journal of Law and Economics. 1960;3:1–44.

[3] Mankiw G. The Pigou Club Manifesto. Available from http://gregmankiw.blogspot.it/2006/10/pigou-club-manifesto.html [Accessed: 2015-10-10].

[4] Pigou AC. The Economics of Welfare. 4th ed. London: McMillan & Co; 1932.

[5] Lewis T. Protecting the environment when costs and benefits are privately known. RAND Journal of Economics. 1996;27:819–47.

[6] Cropper ML, Oates WE. Environmental economics: A survey. Journal of Economic Literature. 1992;30:675–740.

[7]   Porrini D. Environmental policies choice as an issue of informational efficiency. In: Backhaus JG, editor. The Elgar Companion to Law and Economics. 2nd ed. Cheltenham: Edwar Elgar; 2005. p. 350–63.

[8]   Calabresi G. The Cost of Accident. New Haven, CT: Yale University Press; 1970.

[9]   Shavell S. The Economic Analysis of Accident Law. Cambridge, MA: Harvard University Press; 1987.

[10] Kaplow L, Shavell S. On the superiority of corrective taxes to quantity. American Law and Economics Review. 2002;4:1–17. DOI:10.1093/aler/4.1.1

[11] Parry WH, Pizer WA. Emissions Trading versus CO_2 Taxes versus Standards. Resources for the Future, Issue Brief. 2007;5. p. 80-6.

[12] Faure M. Climate Change Liability. Cheltenham: Edwar Elgar Pub; 2011.

[13] Charpentier A. Insurability of Climate Risks. The Geneva Papers of Risk and Insurance. 2008;33:91–104. DOI:10.1057/palgrave.gpp.2510155

[14] Porrini D, Schwartze R. Insurance models and European climate change policies: An assessment. European Journal of Law and Economics. 2014;38:7–28. DOI:10.1007/s10657-012-9376-6

[15] Mills E, Lecomte E. From Risk to Opportunity: How Insurers Can Proactively and Profitably Manage Climate Change. Boston, MA: Ceres Report; 2006.

[16] Agrawala S, Fankhauser S. Economics Aspects of Adaptation to Climate Change. Paris: OECD; 2008.

[17] Munich Climate Insurance Initiative (MCII). Insurance solutions in the context of climate change-related loss and damage. Policy Brief. 2012;6. p. 15.

[18] Kaul I, Conceicao P, Le Goulven K, Mendoza RU. How to improve the provision of global public goods. In: UNDP. Providing Global Goods – Managing Globalization. Oxford: Oxford University Press; 2003.

[19] Nordhaus DW. To tax or not to tax: Alternative approach to slowing global warming. Review of Environmental Economics and Policy. 2007;1:26–40.

[20] Nordhaus DW. After Kyoto: Alternative Mechanisms to Control Global Warming. FPIF Discussion Paper; Washington, DC: Institute for Policy Studies2006.

[21] Metcalf GE, Weisbach D. Linking Policies When Tastes Differ: Global Climate Policy in a Heterogeneous World. Harvard: Harvard Kennedy School. Discussion Paper; 2010. p. 10–38.

[22] Jaffe J, Ranson M, Stavins R. Linking tradable permit system: A key element of emerging international climate change policy architecture. Ecology Law Quarterly. 2009;36:789–808.

9

The Changing Landscape of Energy Management in Manufacturing

Elliot Woolley, Yingying Seow, Jorge Arinez and Shahin Rahimifard

Additional information is available at the end of the chapter

Abstract

The production and use of energy accounts for around 60% of global greenhouse gas (GHG) emissions, providing an intrinsic link between cause and effect. Considering that the manufacturing industry is responsible for roughly one-third of the global energy demand enforces the need to ensure that the manufacturing sector continually strives to reduce its reliance on energy and thus minimise GHG released into the atmosphere. Consequently, efficient management of energy consumption is of paramount importance for modern manufacturing businesses due to well-documented negative impacts regarding energy generation from fossil fuels and rapidly rising worldwide energy costs. This has resulted in a proliferation of research in this area which has considered improvements in energy consuming activities at the enterprise, facility, cell, machine and turret levels. However, there is now a need to go beyond incremental energy efficiency improvements and take more radical approaches to reduce energy consumption. It is argued that the largest energy reduction improvements can be achieved through better design of production systems or by adopting new business strategies that reduce the reliance of manufacturing businesses on resource consumption. This chapter initially provides a review of research in energy management (EM) at various manufacturing focus levels. The inappropriateness of current methods to cater for transformative and radical energy reduction approaches is discussed. In particular, limitations are found at the business strategy level since no technique exists to consider the input of these high level decisions on energy consumption. The main part of the chapter identifies areas of further opportunity in energy management research, and describes a method to facilitate further reductions in energy use and GHG production in manufacturing at the business strategy level.

Keywords: Energy management, Greenhouse gases, business strategy, manufacturing, sustainability

1. Introduction

There are two facts about the future of energy that we know: we will not be able to generate the same quantities from easily accessible fossil-based sources as we currently do, and in the short term this shortfall in energy supply will not be met by 'renewable' sources based on current projections of investment and development [1]. In addition, energy consumption is currently increasing (see Figure 1) and is expected to increase by 22% by 2020 compared to 2011 [2], due partly to increases in demand from China and India [3]. These factors will create what has been termed the 'energy gap' [4]: the difference between demand for energy and the ability to supply this demand, although it should be noted that demand is influenced by supply. The precise magnitude of this energy gap is difficult to predict but it will have a severe influence on the way energy is consumed in the foreseeable future.

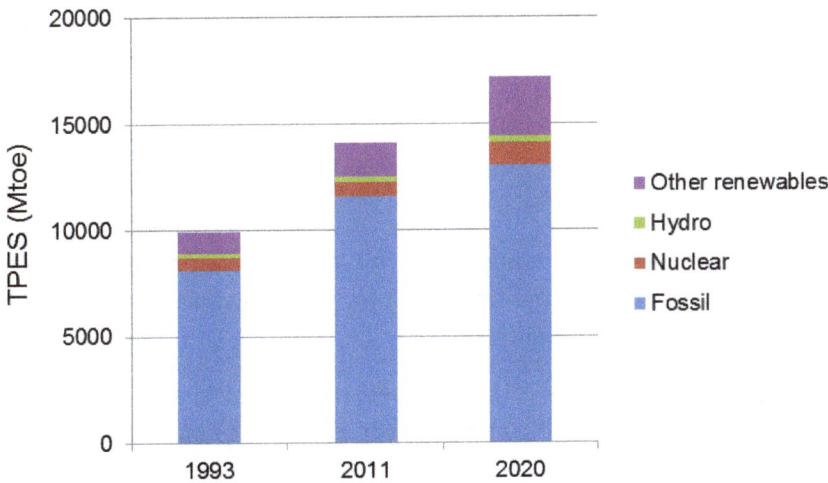

Figure 1. Total primary energy supply of resource by 1993, 2011 and 2020 (data from [2])

The management of energy consumption within the manufacturing sector is particularly important since it is one of the largest energy consuming sectors, directly and indirectly responsible for one-third of global energy use and carbon emissions [5]. This high level of manufacturing related energy consumption is particularly true in developed and developing countries [6]. Energy security is therefore vital for the future of the manufacturing industry and provides a significant incentive to reduce consumption levels. Other incentives also exist in the form of rising energy costs, national and international legislation and consumer demand for greener products. This need has not gone unnoticed and a wide range of approaches have been implemented by companies to reduce reliance on non-renewable (fossil-based) energy sources by improving management practices of energy consuming activities.

These EM techniques have been, in general, quite successful for specific applications but are limited in their scope and so can only ever have a predetermined impact on an enterprise's energy performance. Historically research and the resulting EM techniques have focused on current manufacturing operations and management practices and have therefore tackled problems that can be solved by existing industrial companies. However, modern manufacturing businesses are under ever-increasing pressures to deliver innovative solutions for highly complex tasks for adaptability, economic performance, maintainability, reliability, and scalability. The "Factory of the Future" [7] has to be adaptable not only to the needs of the market but also to the growing requirements for economic and ecological efficiency. Furthermore, such factories will have to take into consideration increased levels of social responsibility and, in particular, environmental sustainability. Based on these challenges, the need for development and validation of new industrial models and strategies is relevant for industrial transformation. These competitive sustainable manufacturing models and strategies will have to aim at achieving long-term economic sustainability through an increase in added value and improved production capability, responsiveness and quality as well as environmental sustainability through the decrease in the consumption of raw materials, water and energy.

The shape of manufacturing is therefore changing as life cycle approaches become more important for impact assessment and the use of ecological data to influence process planning becomes necessary to meet the environmental performance characteristics demanded by government, industry and consumers. In addition new business-strategies[1] are being explored [8–9], with a wide spectrum of product service systems, remanufacturing and product upgradeability, all likely standard models for the future of sustainable manufacturing industries.

This chapter, which identifies the existing trends in EM research for the manufacturing sector and develops new EM regimes that are important for continued advances in energy rationalisation, is divided into three main sections:

- A brief review and analysis of existing EM techniques for manufacturing with a short discussion on developing new EM techniques for evolving manufacturing approaches;

- A more detailed discussion regarding the need for the consideration of business strategies within EM for manufacturing, with the development of a procedure to facilitate this consideration; and finally,

- A case study to demonstrate the applicability of the developed EM procedure for the business strategy manufacturing level.

2. Review of Existing EM Techniques

Manufacturing enterprises and facilities can be highly complex, where monitoring energy consumption and associated GHG production can be cumbersome and expensive. Using a

---

1 In this work, business strategy defines the approach in which a manufacturer takes to fulfil the need or want of the customer. The manufacturer will supply a solution through either the provision of a product or service (which uses a product) and can, therefore, be a short-term or long-term interaction with the customer. Similarly, the business strategy also defines where and when and how often manufacturing processes are undertaken. In short, business strategy defines the model by which a company seeks to generate profit from its customers, but which is also linked to its suppliers and external governing bodies.

well-structured framework can help industrialists identify where to focus their efforts to achieve maximum energy savings. One structured approach to analysing a manufacturing system is to decompose the system hierarchically. A variation of the 'shop floor production model' as developed by the International Organisation for Standardization (ISO) can be used to categorise research conducted on various levels. The adapted model has five levels, ranging from a high level to a specific scope. The levels and the energy considerations for each are summarised as follows:

1. Enterprise level – supply chain of materials or components, network of production sites, inventory hubs, sales and distribution centres, R&D and the integration of various plants.

2. Facility level – building envelope, heating, ventilation and air-conditioning (HVAC), infrastructure of the facility and site energy generation.

3. Production/machine cell level – planning, production engineering and management, supply of material resources and maintenance.

4. Machine level – operation and control of equipment, lighting, cooling, work done on material and communication systems.

5. Turret or tool-chip level – actual transformation of material.

Further, Vijayaraghavan and Dornfeld [10] also suggested that at each level of analysis, there is a corresponding temporal scale of decision making that ranges from several days at the enterprise level to micro-seconds at the tool-chip level. The range of variation in the analysis and temporal scales along with the types of decisions that are made at each level is shown in Figure 2. It is considered by the authors that although the temporal timescales suggested by Vijayaraghavan and Dornfeld [10] are suitable for rapid, high-volume manufacturing processes (e.g., low-tech electronics), for products that are highly complex (e.g., jet engines) or for products that have very long product runs (e.g., cars), these decision timescales need to be extended. This is particularly the case when a decision is based on the use of suffi-cient historical data (which needs to be collected and interpreted). The current research considers temporal timescales between minutes (real-time) and years (strategic).

The following sections use the structure described above to briefly review some of the techniques for EM and minimisation published in recent academic literature. The review is not intended to be comprehensive (see [11] for a review with a wider scope) but is intended to show the shape of research in this field with respect to decision time scales for each manufacturing level. As will be shown, existing research (in EM) falls into the manufacturing level-timescale relationship described by Vijayaraghavan and Dornfeld [10]. This research is concerned with identifying the opportunities and needs for EM techniques that lie outside of this existing envelope. These new areas of energy consideration in manufacturing are shown in Figure 3.

168   Recent Research in Greenhouse Gases

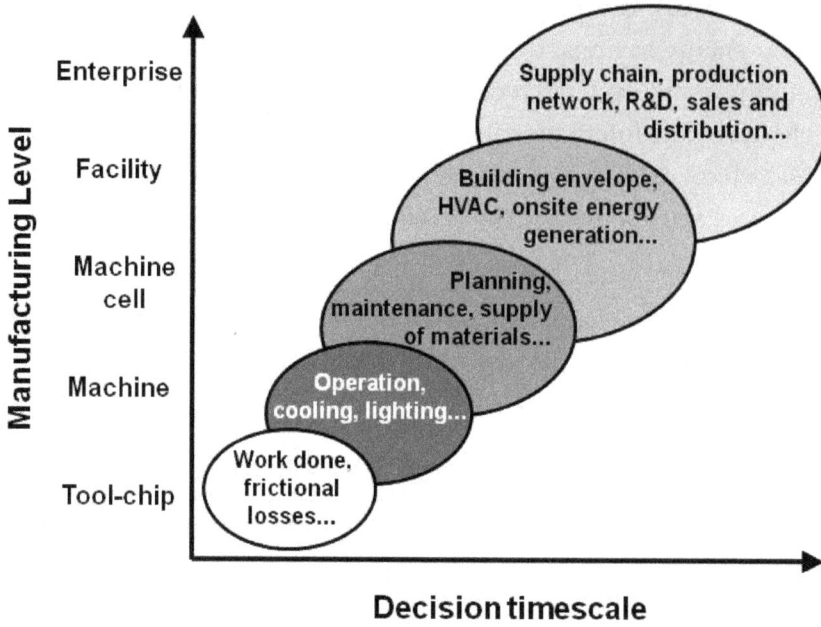

Figure 2. Energy considerations at different manufacturing levels. Adapted from [10].

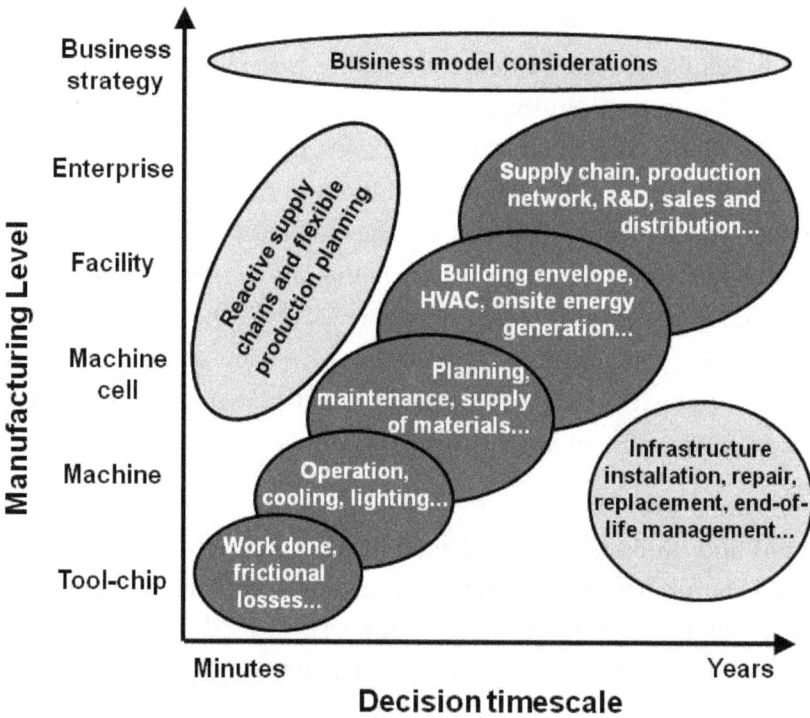

Figure 3. Energy considerations in manufacturing. Existing research predominantly in dark grey areas. New areas described in this research shown in light grey.

2.1. Energy Research on an Enterprise Level

Manufacturing enterprises extend beyond the walls of a factory that just produces goods; they encompass a range of activities from supply chain of materials or components to manufacturing processes and the logistics of the finished product. This involves a network of production sites, suppliers, inventory hubs as well as sales and distribution centres.

Various studies have reported techniques for EM at the enterprise level. Concentrating on logistics Kara and Manmek [12] found that the embodied energy of products could be reduced by selecting local suppliers and avoiding road transport for high quantities of raw materials over long distances. Their model focused on energy, materials and emissions, and waste with considerations for how each of these are used or produced within lengthy supply chains. Supplier location was shown to be a significant factor that can increase or reduce the embodied energy of the raw materials. A similar study [13] used Google Maps to carefully plan and optimise the embodied energy of transportation at the enterprise level. Both approaches require detailed data regarding transport modes and routes and, in the event of instigating changes, may require extensive replanning of multiple cross-linked supply chains. In addition, Kara et al. [14] detail a methodology for assessing the impact of global manufacturing on the embodied energy of products. They studied six different products manufactured from various raw materials in a global manufacturing network and found that product, material and key supply chain parameters played a crucial role.

Other research on the enterprise level has identified that energy improvements can be obtained by changing manufacturing models (e.g., Seliger et al. [15] showed that a phone that is remanufactured consumes less energy than a phone that has been sent to a land fill, over the production, use and end-of-life phases). This is because the remanufacturing pathway, despite requiring energy input into the reverse logistics, avoids repeating manufacturing steps with characteristically high energy consumption and environmental emissions.

The globalisation of businesses has led to long and multi-tiered supply chains, making the introduction of improvements across the entire enterprises complex and difficult. This has been reflected in the number of studies that have been carried out at the enterprise level with most publications focussing on case studies and observed trends rather than on new methodologies [11]. In general, the higher costs, coordination effort and complexity and communication difficulties of implementing sustainable supply chains has led companies to focus on internal activities that present far more achievable environmental (and financial) gains over shorter time periods.

2.2. Energy Research on a Facility Level

Research on the facility level primarily focuses on modelling and reducing the energy consumed by infrastructure and other high level services such as ventilation, lighting, heating and cooling. On-site energy generation is also taken into account.

A review of potential energy savings of a typical manufacturing facility has been performed [16] and focused on high-level redesign strategies. It was concerned with the potential energy saving that can be achieved through optimised building shape and form, improved building envelopes, improved efficiencies of individual energy using devices, alternative energy using systems in buildings, and through enlightened occupant behaviour and operation of building systems. In addition, a method for measuring plant-wide industrial energy savings that takes into account changing weather and production between the pre- and post-retrofit periods has been presented [17].

As a barrier to EM at the facility level, it has been highlighted there is a distinct lack of manufacturing energy performance indicators (EPIs), and this has led to difficulties of modelling 'plant level' energy consumption [18]. Benchmarking energy is essential for EM program development, yet it has been noted that most industries have not, or at least have not been able to, benchmark energy use across their plants. Combining the American Energy Star performance rating system with EPIs, it has been possible to quantify the average energy consumed for the manufacture of best practice vehicles [18]. On a more generic level, the development of energy performance benchmarks and building energy ratings for non-domestic buildings have been reported [19]. They outlined a methodology to develop energy benchmarks and rating systems starting from the very first step of data collection from the building stock.

Finally, on the facility level an economic comparison of three cogeneration steam systems for a wood pulping mill was carried out [20], finding that economic and environmental optimisation could not be achieved simultaneously.

2.3. Energy Research at the Machine Cell Level

At this level, research focuses on planning, production engineering and management, supply of material resources, transport waste material processing and maintenance. Energy flows are closely related to the running of these activities that may be affected by production plans, scheduling times and parameters.

Much of the research reported on the production level involves process planning and process routing for improved energy performance, although most research focuses on costs and cycle times. There is a lack of tools for optimising process flows based on sustainable development objectives (environmental), and those that have been proposed have few practical results [21]. In an attempt to bridge this gap, Tan et al. [21] combined manufacturing process planning and environmental impact assessments using a checklist analysis. They proposed an optimal decision making method for new components that include energy consumption as part of the sustainable development evaluation.

In addition, He et al. [22] have developed green manufacturing process planning and support systems where the raw materials, secondary materials and energy consumption, and other environmental impacts of process planning were optimised. This was supported with databases and model repositories. Integration of the optimisation of energy consumption of processes as part of the process selection algorithm in a process planning program is also possible as and has been demonstrated [23].

Information is critical on a production level: Chiotellis et al. [24], Müller and Löffler [25] and Herrmann et al. [26] have all proposed various information formats to aggregate energy values for decision making on a production level. These groups specifically noted a current lack of monitoring of energy flows within factories. In addition to the lack of monitoring systems, the amount of information required can be very complex and requires a robust framework to deal with information on all levels. They suggest that having online monitoring of the energy consumption within a factory not only provides greater energy transparency, but also provides a stream of useful information to be used for maintenance repair and overhaul. To facilitate this, they have introduced the concept of EnergyBlocks, which can help planners to evaluate the energy consumption profile of various alternatives and to deduce optimal system configuration. However, data volumes increase (almost exponentially) as you move down the manufacturing levels, and it is therefore important to set the correct resolution through appropriate hardware and software systems.

Muller and Loffler's [25] approach to the same problem provides guidance on energy-related decision making during the planning procedure, from the product definition to energy monitoring of the implemented plant. The availability of energy-related data in industry during the planning process is still very rare, and so the main challenge is the development of energy data standards for life cycle engineering (LCE) tools. They have suggested the development of energy performance ratios to influence more detailed standards and instruments such as the dynamic simulation of energy demands.

The correlation of energy usage with operations being performed in the manufacturing system through event stream processing techniques has been successfully implemented [10]. The framework temporally analyses the energy consumption and operational data of machine tools and other manufacturing equipment to enable decision making to improve the environmental performance of the machine tools.

2.4. Energy Research Associated with Production Machines and Equipment

Research associated with the machine level has been concentrated in two subcategories: the energy consumption of the machine for the 'work done' processes and the energy requirement of the machine for auxiliary processes (e.g., cooling and control).

Cooperative Effort on Modelling Process Emissions in Manufacturing (CO2PE!) is an international initiative [27] to cluster forces in different continents, involving machine builders as well as academics, to analyse existing and emerging manufacturing processes for their ecological impact in terms of direct and indirect emissions. Substantial research has been targeted to document, analyse and reduce process emissions for a wide range of available and emerging manufacturing processes [24, 28–31].

In the life cycle phases of product manufacturing, the focus of resource efficiency moves from the material applied per unit to resources used in the various production phases, for example, cooling lubricants, compressed air or hydraulic oil and on the energy requirements of the production processes [24]. Process relevant information is based on equipment energy consumption curves. Each curve is specific to a production equipment item and enables an accurate determination of the energy consumption of the item over the production time.

Similarly, Overcash et al. [32] produced an engineering rule-of-practice-based analysis of separate unit processes used in manufacturing. The information is collated in the form of a unit process life cycle inventory, which then helps to evaluate the manufactured products through the quantification of various parameters, including input materials, energy requirements, material losses and machine variables.

In the context of an integrated consideration of economic and ecological impact, energy profiles are an important basis for deriving optimisations to improve sustainability in manufacturing [33]. On the process level, these profiles permit the identification of substantial energy drivers in machines. In addition, the process-specific energy assessment has taken a step further to develop generalised 'equipment-level' energy models, using average energy intensities of different manufacturing processes to evaluate the efficiency of processing lines [34]. They concluded that modern processes enable smaller dimensions and scales to be produced with larger specific electrical energy requirements. They indicated that energy requirements depend on the production rate and are consequently not constant as assumed by Life Cycle Assessment software packages like Simapro or Umberto.

Dahmus and Gutowski [35] tracked energy flows when characterising the environmental impact of machining, making a distinction between the energy required for chip formation and operating the manufacturing equipment (Figure 4). In their studies, they showed that machine tools with increasing levels of automation reveal higher basic energy consumptions that result from the amount of additional integrated machine components.

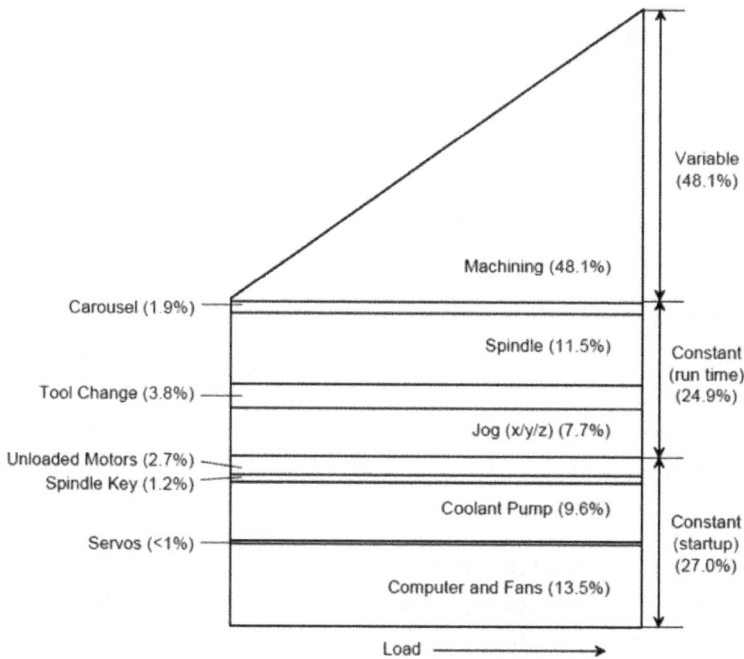

Figure 4. Machining energy use breakdown for a 1988 Cincinnati Milacron automated milling machine with a 6.0 kW spindle motor [35].

More specifically, a study on the energy consumption of cutting found that high speed cutting required less energy per unit manufactured compared to the conventional cutting speed [36]. They also found that the installation of kinetic energy recovery systems (KERS) can reduce average power consumption by up to 25% depending on workpiece geometry and machining time. As the energy efficiency of the system is highly part specific, a KERS should be custom defined. A framework has also been developed [37] for the recovery of waste heat energy from manufacturing processes.

Along the same theme, the improvement potential in two types of manufacturing equipment for discrete part production has been discussed [29]. Power requirements for activities in a machine tool was investigated and classified into productive and non-productive periods.

In contrast, however, Fleschutz et al. [38] conducted an energy simulation on 12 similar industrial robots within a workstation and found that the assigned operations strongly influenced the energy consumption of the respective robot. Even though the operating hours are the same for the robots, those that had more kinematic movements and little idle time resulted in energy consumptions that were double the other robots.

Process conditions and energy consumption are not normally static but depend directly on the specific conditions of the process and/or the production setting. The initiation of energy labels for production machines indicating the amount of electrical energy consumed for various production processes has been proposed [30]. Such information can be estimated by summarizing the electrical energy consumptions of single machine components (pumps and engines) or by using energy profiles to reveal the holistic energy absorption that is needed by machine tools.

There are, however, significant challenges in obtaining sufficient energy consumption data. It has been proposed that in the future, both the manufacturer of equipment and the operator should use consistent parameters to describe the energy performance of manufacturing systems and that equipment should have standardised metering ports [39].

2.5. Energy Research on the Turret Level (Theoretical Process Energy)

The lowest and most focused manufacturing level is the turret level, which represents the actual material transformation process itself. Energy assessment and management at this level involves knowledge of the interactions of the mechanical and chemical processes in order to establish theoretical energy consumption values of the process.

The research and application of improvement in energy efficiency at the turret level is highly process specific and therefore less appropriate to general application. As a reflection of this, less research has been published at this level. Most research is evaluative with little scope provided for developing models. Consequently, only a few examples reviewed here are used in further analysis in this chapter.

At this focus level, it has been shown that in machining, the ideal process energy is independent of operating parameters such as tool speed, feed and depth of cut [40]. Instead the machining energy is dependent on setup parameters, such as choice of cutting fluid, tool rake angle and part design parameters (material selection and the volume of material removed). Draganescu et al. [41] conducted experiments to model machine tool efficiency so that the specific consumed energy could be determined for establishing cutting parameters and the consumed energy necessary for removing a certain quantity of chips. Amongst many examples of theoretical mathematical modelling of machine processes, Draganescu et al. [41] and Kalpakjian and Schmid [42] have looked at the specific energy consumption for milling, and Ghosh et al. [43] have modelled the specific energy requirement of deep grinding.

Other studies of theoretical energy consumption of other manufacturing process can be found in [42] who give detailed explanations and descriptions of the energy required for cutting, forming and deformation. A detailed analysis on the specific cutting energy for bandsawing different work piece materials has been carried out [44]. The minimal energy required for turning and the optimal conditions for machining a product has been studied [45] and finally Kuzman and Peklenik [46] have done an energy evaluation of cold forming processes.

3. An Overview of the Current Scope of EM

There are clearly a large number of EM techniques that manufacturers can implement in order to reduce their energy consumption and their generation of GHGs. These techniques focus on many different aspects, and it can be confusing for manufacturers to decide which approaches are best for their particular setup. One way of categorising these different EM techniques is to define the temporal decision timescale. Within manufacturing, EM decisions can be made on one of four timescales: real-time (minutes-hours), operational (days-weeks), tactical (months) and strategic (years), with EM techniques at these levels being implemented by different groups of people who operate at different management levels.

Positioning the EM techniques reviewed in this chapter into a research map (Figure 5) that has manufacturing level and decision timescale as its axes reveals almost intuitive results. The more focused (lower manufacturing level) an EM technique, the shorter the timescale on which decisions can be made: adjusting machine setup parameters (turret level) can be done by one person in a few minutes, whereas reconfiguring a supply chain (enterprise level) will take a team of people months or even years. In the research map, this correlation seems linear, but since the x-axis is not continuous and the y-axis is not quantitatively scaled, a strict correlation is undefinable and inappropriate. Nonetheless, there are clear areas of the map that are not occupied by any of the reviewed EM techniques, and it is therefore suggested that there is a need for research to be undertaken to address these areas.

There are two areas of research in the current map (Figure 5) that would benefit from a growth research: lifecycle process system planning and eco-intelligent manufacturing and agile supply chains. A further area of research, not currently represented on the above map, is required to look at the impact on energy consumption of existing and future business strategies. These three research opportunities are discussed in the following sections, with EM at the business strategy level given a more in-depth consideration due to the importance of its potential for limiting GHG production (energy consumption) in order to meet a specific customer need.

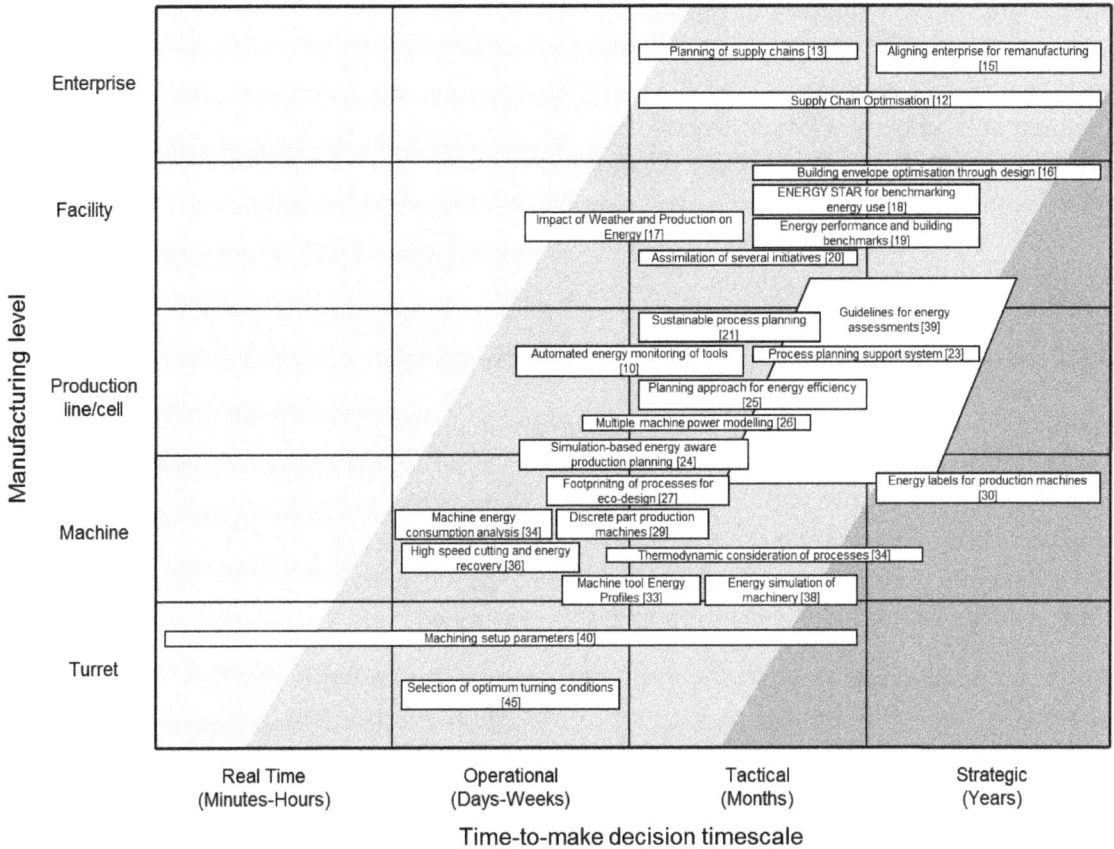

Figure 5. Research map showing the relation between manufacturing level and associated decision timescale. The mid-grey region is heavily populated, whereas EM techniques at for short-term facility and enterprise energy consumption and long-term process level energy consumption are largely undeveloped.

3.1. Lifecycle Process System Planning

Recently, as with products, manufacturers are beginning to take a life cycle view of not just their factories, but also the processes within the factories [47]. Primarily for economic reasons but also from an environmental point of view they are beginning to consider how best to ensure their machinery is maintainable and upgradeable, and what will happen to it at its end of life. Such a task can be highly complex since it is difficult to predict the requirement of future or long-term process capabilities, process utilisation levels and also production floor layouts. Life cycle process planning is, therefore, heavily dependent on a company's ability to roadmap the sector it operates in. However from an environmental perspective, it is highly important to be able to consider the resource intensity of processes throughout their lifetime. A host of life cycle process planning tools are therefore required to assist manufacturers in managing their long-term process requirements. For this, they require the development of an assessment framework and decision support tool to understand the life cycle impact of individual processes and process chains so that strategic decisions can be made about the purchase or upgrade of machinery to ensure minimal environmental and economic impact.

Such a suite of tools requires a consideration of energy consumption to ensure the processes are able to integrate with the long-term energy supply strategy of a particular facility. Life cycle process EM should consider flexible, reconfigurable process chains, peripheral energy requirements (cooling, transport systems, etc) and factory layout to ensure the most appropriate planning is carried out. EM tools developed in this area will need to provide manufacturers with clear strategies for the minimisation of process energy consumption over periods of many years. Such considerations will be key to the development of eco-efficient factories for sustainable industrial systems.

3.2. Eco-Intelligent Manufacturing and Agile Supply Chains

At the short end of the temporal decision timescale on the research map, there currently exists another area where no or very few EM techniques exist. In this region, which covers manufacturing levels between cell and enterprise, manufacturers are beginning to consider new short-term influences on their production and operations processes to improve economic and environmental performance. As part of a holistic approach, they require techniques to help them manage their energy consumption at this timescale.

At the production cell and factory levels, real-time consideration is required to be sensitive to energy availability, which may come from several sources (almost certainly involving traditional fossil-based energy) and which for any renewable component may vary in supply. In such circumstances, it is necessary to be able to rapidly influence production scheduling to ensure that energy intensive processes are carried out at periods of higher renewable energy availability to maximise the environmental benefit from this type of energy. Smart metering and smart energy grids are required to influence this eco-intelligent manufacturing. In fact, a new production planning regime, 'environmental resource planning', is required to not just take into account the immediate availability of energy supply mix, but a full range of eco-indicators, including emissions, water consumption, idle time of processes, and staff availability.

There is also a need for industry to be able to consider and manage energy at the enterprise level, but on the short real-time and operational timescales. Supply chains have historically been set up to optimise for time and cost to give manufacturers the best possible competitive advantage. However, this approach has led to disadvantages in fluctuating markets and increasingly, manufacturers are seeking to remain resilient by creating flexible reactive supply chains. The Triple-A Supply Chain described by Lee [48] promotes the need for agile supply chain arrangements that are able to respond to short-term changes in supply, allowing rapidly changing consumer demands or unforeseen disruptive events to be more easily worked around. The energy implications in an agile supply chain need to be managed even if it might not be a primary consideration. It is important for companies to be able to account for the resources that are required for their products from a life cycle approach, which includes being able to optimise the supply chain. Methods for systematically reacting to supply and demand problems are being developed [48] and incorporated into everyday business practices. EM techniques that are able to consider and influence these short-term reactive changes to supply chain operation will be essential for improving and maintaining environmental performance in the future manufacturing industry.

3.3. Energy Management at the Business Strategy Level

The preceding reviewed and hypothesised EM techniques for manufacturing are suitable for current make-sell business models that are optimised for economic benefit. Since profit is generated from the sale of products, successful manufacturing businesses have historically been those that produce and sell more than their competitors: a culture that has led to a disregard for resource consumption and pollution levels. This practice is now changing as businesses and consumers become more aware of, and active in achieving long-term sustainability. Manufacturers are investigating and implementing radical new strategies to remain commercially competitive whilst reducing resource consumption. It is therefore proposed here that new EM techniques will be required to support these new manufacturing business strategies.

The field of manufacturing has seen many improvements in sustainability performance over the last few decades. Strategies to reduce waste, lower emissions, improve energy efficiency, and so on have been implemented across the board but such activities have largely made improvements for individual processes only. In the long term, industrial sustainability will not be achieved simply by the development of new technologies or the utilisation of iterative improvements of current production processes. Accordingly, the appropriateness of existing business models are being challenged [49–51] for a future of industrial sustainability. The configuration of the industrial system will evolve dramatically, introducing new concepts such as cradle-to-cradle [52], slow manufacturing, local manufacturing [53], product service systems [9], and product compatibility and upgradeability [54].

An additional manufacturing level is required to consider implications of cost, energy and other resource consumption for these new business strategies [55]. Since for manufacturers the primary consideration in choosing a business strategy will always come down to cost and profit, it is unlikely that any manufacturing activities will be fully optimised for energy efficiency. However, by being able to understand how energy can be considered, measured and managed at this high level, significant energy improvements could still be made. Through the development of new EM techniques, it is important to be able to consider the life cycle manufacturing energy consumption of products as well as the life cycle energy consumption of manufacturing processes and facilities used to produce these products. Figure 6 shows an updated and simplified version of the EM technique research map, which includes the proposed business strategy manufacturing level. Preliminary guidelines for the consideration of energy consumption factors at this business strategy level are discussed by comparing two different business strategies for furniture.

Two distinct and simple business strategies for manufacturing household furniture may be described as the provision of either low-cost, short lifespan, or expensive, long lifespan products. The decision on which strategy to take is made at the conception stage of the business and may depend upon existing supply chain links, market opportunities or available workforce skill set for example. It is unlikely that such a strategic decision would ever be made

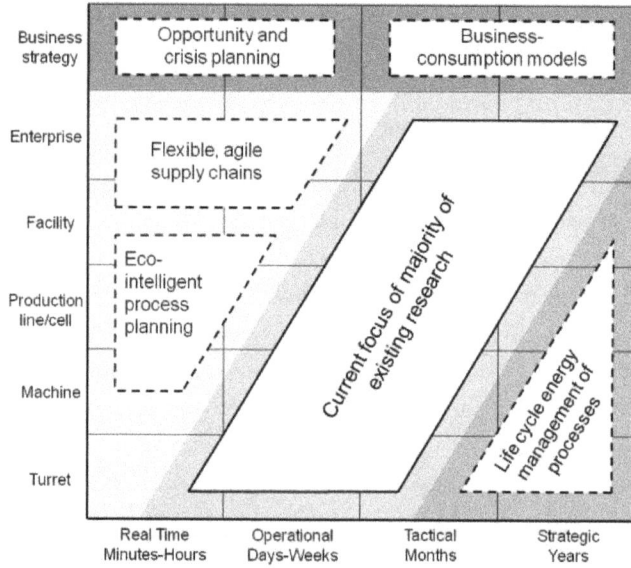

Figure 6. Manufacturing level-decision timescale map showing positioning of required focus for EM to contribute towards industrial sustainability.

purely on the merits of energy consumption levels, but the decision itself has a significant implication on energy use. Table 1 shows some differences between the individual approaches for the manufacture of the furniture. EM techniques are required to assess the different impacts of the two strategies to enable effective EM decisions to be made.

For the short life products, the manufacturer requires a rapid production throughput, and low cost, lightweight materials are likely to be used, which can be quickly manufactured. Because of the highly competitive low selling price of the short-life goods, profit margins are low and therefore there is an economic need to have high volumes of production ensuring that the market remains in constant need of new furniture. For this type of production, low manufacturing costs are essential and so rapid, highly efficient and centralised manufacture is of key importance, and therefore, the embedded energy [56] per product is likely to be relatively low.

Short-life furniture	Long-life furniture
· Energy efficient process	· High quality processes
· Rapid production	· Quality-driven production
· Automated production	· Semi-automated production
· High throughput	· Low throughput
· Low profit margins	· High profit margins

Table 1. A comparison of different manufacturing considerations for short- and long-life furniture.

Conversely, for long life products, materials and manufacturing processes are quality driven with less regard for cost, which will be easily recuperated by high profit margin sale prices. As the products are intended to last a long time, the manufacturer is keen on brand awareness and will ensure their products are of the highest quality. Slower, more energy intensive procedures are likely to be used with additional finishing, inspection and testing processes. Distributed manufacture from small facilities is likely to be preferred to be adaptable to regional market requirements. For this type of product, the embedded energy per unit is likely to be relatively high.

Since the object of manufacturing is to provide a functional unit for a particular market need, a key indicator is the length of time a product fulfils a need. The longer the life of the product with respect to energy consumed, the more energy efficient the manufacturing of the product. This energy per product year, E_{PY}, can be expressed in simple terms as E_{PY} = energy required for manufacture/product lifetime.

Applying this measure to the present example we can consider that the total energy consumption required for an item of furniture is the summation of embedded product energy, E_E, and amortised (i.e., per product) energy from the lifetime of the production processes (i.e., the energy required to produce, maintain and repair the process machinery), E_P, and the amortised energy from the facility, E_F, then the energy attributed to each year of the product's life can be expressed as $E_{PY} = (E_E + E_P + E_F)/L_P$, where L_P is the average anticipated product life in years. Note that if the producer has responsibility for the disposal of the product at its end of life, the energy requirement for this also needs to be considered in the above equation. Other considerations at this business strategy level might include the possibility of remanufacture (lowering future E_E), availability of renewable energy (lowering the impact of E_E), production at low energy demand periods and maintenance contracts.

The product lifetime is of utmost importance in the case where significantly different lifetimes of products are being considered. In this example of furniture, it is not unreasonable to assume an order of magnitude difference in lifespan. Assuming there is no significant difference between the production energy, the business strategy that manufactures longer life furniture will be preferable.

Regardless of which business strategy is used, the process of considering the energy consumption factors throughout the life cycle of the products can be used to influence the selection of manufacturing processes, facilities, facility operation times, and so on. Clearly, as with EM techniques at lower manufacturing levels, approaches at the business strategy level require availability of suitably reliable or indicative data, or appropriate assumptions. In addition because of the complexity of different business models, such high level EM may require significant input from techniques focussing at lower levels of EM (e.g., HVAC control to reduce E_E) to yield the best results.

There is a need for the development of new EM techniques at the business strategy level that assist manufacturers not only in deciding which production models are least energy intensive, but also in minimising energy consumption at the highest level with an integrated approach. Importantly, such EM techniques need data for the specific stages in a product's life cycle for which the manufacturer is responsible. By using energy per product year as the unit of measure, it is possible to compare between dissimilar manufacturing and business approaches.

The general approach suggested here is a three stage consideration of energy at the business strategy level. In the first stage, it is essential to set the boundaries and contributing factors for energy consumption. Boundaries will include everything that a manufacturer is responsible for or has direct control over during the entire product/s life time and could extend to the supply chain, providing sufficient information is available. Contributing factors should focus on high level energy consuming tasks such as embodied product energy, impact of energy sources, embodied life cycle energy of the factory infrastructure, post-life product responsibility and life span of manufactured product.

The second stage in the approach requires an understanding of the relationship between different energy considerations, assignment of appropriate variables and development of any relationships between factors. In most cases, it will be useful to evaluate the total energy consumption per product year for which the manufacturer is directly responsible (i.e., not during use cycle). It is not important to fully understand the details of each energy contributing factor, but it is important to understand how certain factors relate to one another. An example of this is described in Section 4.

The third and final stage is to identify the factor(s) that have the largest energy contribution and make improvements in these areas as appropriate.

Obviously, the current approach is generalised but is intended to give guidance for the development of future EM techniques for decisions at the business strategy level. Of key importance is sufficient understanding of the impact of business strategy on energy consumption and the specific actions that can be taken to reduce this reliance on energy.

4. A Case Study for EM Method at the Business Strategy Level

The life cycle approach for considering energy at the business strategy level can be applied to any product, provided that sufficient consideration is given to the contributing factors. The level of detail can be adjusted to allow for different levels of data availability or understanding, but it is important to assess the need for different energy consumption factors within the life cycle of the product.

It may be criticised that consideration of some factors are difficult at the business strategy planning stage. However, it is not necessarily essential to have firm data in order to be able to assess different strategies; sensible assumptions can be used to evaluate the energy implications of business strategies. The following example compares different business strategies for the manufacturing and provision of steel roofing material using energy data from [12]. Considerations for energy consumption should include manufacture of the sheeting, transport, maintenance or replacement and any energy consumption or benefit from end-of-life management. The data used for this case study is shown in Table 2.

Factor	Values
Steel sheet production[#]	178 MJ/m^2
Transport[#]	2 MJ/m^2
Replacement section manufacture	3.6 MJ/m^2 year
Maintenance of roof	3.6 MJ/m^2 year
End of life management[#]	-48 MJ/m^2

[#]Data from [12].

Table 2. Numerical values used for comparison between PSS and sale business strategies for steel roofing.

A manufacturer may evaluate the benefit with respect to energy consumption of supplying galvanised steel roof sheeting under a product service system (PSS) basis as opposed to the more common make-sell business model. Under a PSS, the building owner does not actually own the roofing material (ownership remains with the producer) but lease it on a fixed term basis (e.g., for the period of occupation of the building) with maintenance costs being covered by the producer. The 'user' simply pays for the use of the roofing. Setting the boundaries of the comparison between manufacturing processes carried out in-house by the manufacturer and the end-of-life management of the galvanised steel sheeting, Table 3 shows the different factors for consideration in this scenario.

Factor	Sale	PSS	Values/Assumptions
Steel sheet production (MJ/m^2)	$E_{x,prod}$	$E_{y,prod}$	$E_{x,prod}=E_{y,prod}=178$
Transport (MJ/m^2)	$E_{x,tran}$	$E_{y,tran}$	$E_{x,tran}=E_{y,tran}=2$
Replacement section manufacture (MJ/m^2 year)	$E_{x,rep}$	0	$E_{x,rep}=E_{x,prod}/50=3.6$
Maintenance of roof (MJ/m^2 year)	0	$E_{y,mnt}$	$E_{y,mn}=E_{y,prod}/50=3.6$
End-of-life management (MJ/m^2)*	0	$E_{y,eol}$	$E_{y,eol}=-48$
Lifetime (years)	z	kz	z = 20 years

k, the ratio of the life of the PSS roofing to the life of the customer owned roofing.

*If the manufacturer instigates the recycling of the material, they can justify off-setting any energy benefit against their manufacturing energy consumption.

Table 3. Energy consumption factors considered for comparison between PSS and sale business strategies for steel roofing.

Using the factors from Table 3 in the equation $E_{PY}=(E_E+E_P+E_F)/L_P$, the energy per product year can be written for the sale and PSS business strategies, respectively, as $E_{x,py}=((E_{x,prod}+E_{x,tran})/z)+E_{x,rep}$ and $E_{y,py}=((E_{y,prod}+E_{y,tran}+E_{y,eol})/kz)+E_{y,mnt}$.

These two basic expressions can be used to determine the significant contributing factors of both the roofing sale and PSS strategies. Since from the above we can obtain the equality $((E_{x,prod}+E_{x,tran})/z)+E_{x,rep}=((E_{y,prod}+E_{y,tran})/z)+E_{x,mnt}$, then we can obtain, $E_{y,py}=(E_{x,py}/k)+(E_{y,eol}/kz)$.

Using the values from Table 3, and given that it can be shown that $E_{x,py} = 9.8$ MJ/yr/m^2, we find that, $E_{y,py}=(7.4/k)$.

Thus it can be shown that if $k > 0.76$, then it will always be beneficial from a manufacturer's energy perspective to opt for the PSS strategy. In the PSS strategy, since the manufacturer generates income from each year the roofing material is in service, rather than from a one-off income generated through the sale of the steel sheeting, the value of k will likely be greater than 1. The incentive to prolong the life of the product by the manufacturer means that addition care may be taken in the maintenance of the roofing, at the expense of $E_{y,mnt}$.

Based on the given example, to make the sale strategy more competitive in terms of energy, manufactures could focus on methods of extending the life of the roofing (e.g., by additional coatings) or significantly reducing the embedded product energy (which will likely have a negative impact on life expectancy). As no actual energy is consumed at the business strategy level, energy improvements will ultimately come from the implementation of EM techniques at lower manufacturing levels.

In the given case study, it is possible to introduce additional terms into this comparison which may look at energy implications such as process machinery and infrastructure life cycle energy costs, warranty repair, supply chain PSS, for example, depending on the company's scope and business model.

Using the three stage approach proposed in this chapter it is possible to compare between different business strategies using limited data or assumptions. The output will give an indication as to the energy consumption factors that need to be considered in more detail to ensure the overall minimisation of energy use for manufacturing activities. However, significant work is required by academia and the manufacturing industrial sector to develop more focused EM techniques for the business strategy level.

5. Concluding Discussions

The need for EM and rationalisation within manufacturing has led to the development of a very large number of EM techniques. These techniques cover issues from machine–tool interaction to distribution logistics and supply chain management. A review of a cross-section of EM techniques has revealed a correlation between the manufacturing levels and decision timescale. As a consequence, there are areas of research that have not yet been addressed, such as the life cycle evaluation of production processes and the short-term management of energy supply and supply chain operations. This chapter asserts that there is a requirement for EM techniques to be developed in these currently uninhabited research spaces since manufacturers are continually searching for ways to reduce energy costs and improve environmental performance. Considerations for approaches to developing EM for life cycle impact of production processes include flexible, reconfigurable process chains, factory layouts and upgradability and maintainability. For eco-intelligent factories and reactive supply chains, energy supply security, production flexibility and supply chain agility will form the cornerstones of EM for manufacturers of the future.

In addition, as manufacturers consider more radical ways of reducing their environmental impact and maintaining market share in volatile consumer markets, new business strategies are being considered to generate income in ingenious ways. These new strategies are being developed with economic benefit as the primary focus, but it is also important for manufacturers to consider EM of these strategies to minimise the energy consumption of their activities in the long term. Therefore, building on the existing scope of EM techniques which have focussed on the manufacturing levels, this work suggests that new techniques need to be developed for an additional manufacturing level, namely business strategy, which is positioned above the enterprise level in the hierarchy. The business strategy level has been shown to be slightly different from the other manufacturing levels in that no energy is actually consumed at this level, but energy consumption for products and services are largely defined by decisions at this level. It is therefore reasoned that approaches to EM for business strategies should consider the energy expenditure that the manufacturer has direct control over for the life cycle of the product. This approach allows different business models to be evaluated using the same core framework, even if the strategies are markedly different.

A three stage high level procedure has been described for EM at the business strategy level, which consists of definition of scope and energy contributing factors, identification of inter-relationships between energy factors and a comparison of potential strategies and finally a focus on the largest energy consuming factors using techniques from lower manufacturing levels. Of key importance at the business strategy level is that there is no requirement for detailed energy data that may be difficult to obtain at a planning stage. Instead, comparisons may be made between strategies based on a few well-grounded assumptions.

In summary, as manufacturing businesses become more energy aware and seek to remain competitive in highly transient and environmentally focused markets, new business strategies and increased production flexibilities are being explored. The manufacturing industry is evolving for the better, and new EM techniques will be essential in supporting this revolution.

Author details

Elliot Woolley[1*], Yingying Seow[2], Jorge Arinez[3] and Shahin Rahimifard[1]

*Address all correspondence to: e.b.woolley@lboro.ac.uk

1 Centre for Sustainable Manufacturing and Recycling Technologies (SMART), Wolfson School of Mechanical and Manufacturing Engineering, Loughborough University, Loughborough, UK

2 Jacobs, New City Court, London, UK

3 Manufacturing Systems Research Lab, General Motors R&D Center, Michigan, USA

References

[1]   Cowan, K.R., and Daim, T., 2009. Comparative technological road-mapping for renewable energy, *Technology in Society*, 31 (4), 333–341.

[2]   World Energy Council (WEC), 2013. World Energy Resources – 2013 Survey. Available at: https://www.worldenergy.org/wp-content/uploads/2013/09/Complete_WER_2013_Survey.pdf [Accessed 16 October 2015].

[3]   International Energy Agency (IEA), 2009. World Energy Outlook 2009. Available at: http://www.iea.org/country/graphs/weo_2009/fig1-1.jpg [Accessed 9 December 2012].

[4]   alameh, M.G., 2003. Can renewable and unconventional energy sources bridge the global energy gap in the 21st century? *Applied Energy*, 75 (1–2), 33–42.

[5]   Evans, S., Bergendahl, M., Gregory, M., and Ryan, C., 2009. *Towards an Industrial Sustainable System*, Institute for Manufacturing, University of Cambridge.

[6]   Lee, C-C., and Chang, C-P., 2007. Energy consumption and GDP revisited: A panel analysis of developed and developing countries, *Energy Economics*, 29 (6), 1206–1223.

[7]   Jovane, F., Westkämper, E., Williams, D.J., 2009. *Towards Competitive Sustainable Manufacturing*, Springer: Berlin, ISBN 978-3-540-77011-4.

[8]   Linton, D., Klassen, R., and Jayaraman, V., 2007. Sustainable supply chains: An introduction, *Journal of Operations Management*, 25 (6), 1075–1082.

[9]   Tukker, A., 2004. Eight types of product–service system: Eight ways to sustainability? Experiences from SusProNet, *Business Strategy and the Environment*, 13 (4), 246–260.

[10] Vijayaraghavan, A., and Dornfeld, D., 2010. Automated energy monitoring of machine tools, *CIRP Annals – Manufacturing Technology*, 59 (1), 21–24.

[11] Seuring, S., and Muller, M., 2008. From a literature review to a conceptual framework for sustainable supply chain management, *Journal of Cleaner Production*, 16, 1699–1720.

[12] Kara, S., and Manmek, S., 2010. Impact of manufacturing supply chains on the embodied energy of products, *Proceedings of the 43rd International Conference on Manufacturing Systems*, Vienna, Austria, (pp 187-194).

[13] Pearce, J.M., Johnson, S.J., and Grant, G.B., 2007. 3D-mapping optimization of embodied energy of transportation, *Resources, Conservation and Recycling*, 51 (2), 435–453.

[14] Kara, S., Manmek, S., and Herrmann, C., 2010. Global manufacturing and the embodied energy of products, *CIRP Annals – Manufacturing Technology*, 59, 29–32.

[15] Seliger, G., Kernbaum, S., and Zettl, M., 2006. Remanufacturing approaches contributing to sustainable engineering, *Gestao & Producao*, 13, 367–384.

[16] Harvey, D., 2009. Reducing energy use in the buildings sector: Measures, costs, and examples, *Energy Efficiency*, 2, 139–163.

[17] Kissock, K.J., and Eger, C., 2008. Measuring industrial energy savings, *Applied Energy*, 85 (5), 347–361.

[18] Boyd, G., Dutrow, E., and Tunnessen, W., 2008. The evolution of the ENERGY STAR® energy performance indicator for benchmarking industrial plant manufacturing energy use, *Journal of Cleaner Production*, 16, 709–715.

[19] Hernandez, P., Burke, K., and Lewis, J.O., 2008. Development of energy performance benchmarks and building energy ratings for non-domestic buildings: An example for Irish primary schools, *Energy and Buildings*, 40 (3), 249–254.

[20] Cakembergh-Mas, A., Paris, J., and Trépanier, M., 2010. Strategic simulation of the energy management in a Kraft mill, *Energy Conversion and Management*, 51 (5), 988–997.

[21] Tan, X., Liu, F., Dacheng, L., Li, Z., Wang, H., and Zhang, Y., 2006. Improved methods for process routing in enterprise production processes in terms of sustainable development II, *Tsinghua Science & Technology*, 11 (6), 693–700.

[22] He, Y., Liu, F., and Cao, H., 2005. Process planning support system for green manufacturing and its application, *Computer Integrated Manufacturing Systems*, 11 (7), 975–980.

[23] He, Y., Liu, F., Cao, H., and Zhang, H., 2007. Process planning support system for green manufacturing and its application, *Frontiers of Mechanical Engineering in China*, 2 (1), 104–109.

[24] Chiotellis, S., Weinert, N., and Seliger, G., 2010. Simulation-based, energy-aware production planning, *Proceedings of 43rd CIRP International Conference on Manufacturing Systems*, Vienna, Austria, pp. 165–172.

[25] Müller, E., and Löffler, T., 2010. Energy efficiency at manufacturing plants – A planning approach, *Proceedings of 43rd CIRP International Conference on Manufacturing Systems*, Vienna, Austria, pp. 5–12.

[26] Herrmann, C., Bogdanski, G., and Zein, A., 2010. Industrial smart metering – Application of information technology systems to improve energy efficiency in manufacturing, *Proceedings of 43rd CIRP International Conference on Manufacturing Systems*, Vienna, Austria, pp. 134–142.

[27] Duflou, J., 2009. *CO2PE! Cooperative Effort on Process Emissions in Manufacturing*, Manufacturing Technology Platform (MTP).

[28] Pusavec, F., Krajnik, P., and Kopac, J., 2010. Transitioning to sustainable production – Part I: Application on machining technologies, *Journal of Cleaner Production*, 18 (2), 174–184.

[29] Devoldere, T., Dewulf, W., Deprez, W., Willems, B., and Duflou, J.R., 2007. Improvement potential for energy consumption in discrete part production machines, *Proceedings of the 14th CIRP Conference on Life Cycle Engineering, June 11–13, 2007,* Waseda University, Tokyo, Japan, pp. 311–316.

[30] Herrmann, C., Bergmann, L., Thiede, S., and Zein, A., 2007. Energy labels for production machines – An approach to facilitate energy efficiency in production systems, *Proceedings of 40th CIRP International Seminar on Manufacturing Systems,* Liverpool, UK.

[31] Gutowski, T., Dahmus, J., Thiriez, A., Branham, M., and Jones, A., 2007. A thermodynamic characterization of manufacturing processes, *Proceedings of the 2007 IEEE International Symposium on Electronics and the Environment,* Orlando, USA, pp. 137–142.

[32] Overcash, M., Twomey, J., and Kalla, D., 2009. Unit process life cycle inventory for product manufacturing operations, *ASME International Manufacturing Science and Engineering Conference,* West Lafayette, IN, USA.

[33] Herrmann, C., Zein, A., Thiede, S., Bergmann, L., and Bock, R., 2008. Bringing sustainable manufacturing into practice – The machine tool case, *Sustainable Manufacturing VI: Global Conference on Sustainable Product Development and LCE,* Pusan, Korea.

[34] Gutowski, T., Dahmus, J., and Thiriez, A., 2006. Electrical energy requirements for a manufacturing process, *Proceedings of CIRP International Conference on Life Cycle Engineering 2006,* Leuven, Belgium, pp. 623–628.

[35] Dahmus, J., and Gutowski, T., 2004. An environmental analysis of machining, *Proceedings of the 2004 ASME International Mechanical Engineering Congress and RD&D Exposition,* Anaheim, California, USA.

[36] Diaz, N., Helu, M., Jarvis, A., Tönissen, S., Dornfeld, D., and Schlosser, R., 2009. Strategies for minimum energy operation for precision machining, *Proceedings of MTTRF 2009 Annual Meeting,* Shanghai, People's Republic of China.

[37] Luo, Y., Woolley, E., Rahimifard, S., and Simeone, A., 2015. Improving energy efficiency within manufacturing by recovering waste heat energy, *Journal of Thermal Engineering,* 1 (5), 337–334.

[38] Fleschutz, T., Azwan Abdul Rahman, A., Harms, R., and Seliger, G., 2010. Assessment of life cycle impacts and integrated evaluation concept for equipment investment, *17th CIRP International Conference on Life Cycle Engineering,* Zhang, H.C., Liu, Z., and Liu, G., eds., Hefei, China.

[39] Müller, E., and Löffler, T., 2009. Improving energy efficiency in manufacturing plants – Case studies and guidelines, *16th CIRP International Conference on Life Cycle Engineering (LCE 2009)*, Cairo, Egypt, pp. 465–472.

[40] Munoz, A.A., and Sheng, P., 1995. An analytical approach for determining the environmental impact of machining processes, *Journal of Materials Processing Technology*, 53 (3–4), 736–758.

[41] Draganescu, F., Gheorghe, M., and Doicin, C.V., 2003. Models of machine tool efficiency and specific consumed energy, *Journal of Materials Processing Technology*, 141 (1), 9–15.

[42] Kalpakjian, S., and Schmid, S.R., 2008. *Manufacturing Processes for Engineering Materials*, Prentice Hall, Singapore.

[43] Ghosh, S., Chattopadhyay, A.B., and Paul, S., 2008. Modelling of specific energy requirement during high-efficiency deep grinding, *International Journal of Machine Tools and Manufacture*, 48 (11), 1242–1253.

[44] Sarwar, M., Persson, M., Hellbergh, H., and Haider, J., 2009. Measurement of specific cutting energy for evaluating the efficiency of bandsawing different workpiece materials, *International Journal of Machine Tools and Manufacture*, 49 (12–13), 958–965.

[45] Rajemi, M.F., Mativenga, P.T., and Aramcharoen, A., 2010. Sustainable machining: Selection of optimum turning conditions based on minimum energy considerations, *Journal of Cleaner Production*, 18 (10–11), 1059–1065.

[46] Kuzman, K., and Peklenik, J., 1990. Energy evaluation of cold-forming processes, *CIRP Annals – Manufacturing Technology*, 39 (1), 253–256.

[47] Labuschagne, C., and Brent, A., 2004. Sustainable project life cycle management: The need to integrate life cycles in the manufacturing sector, *International Journal of Project Management*, 23 (2), 159–168.

[48] Lee, H.L., 2004. The triple – A supply chain, *Harvard Business Review*, October, 2–11.

[49] Comes, S., and Berniker, L., 2008. Business model innovation. In: D. Pantaleo & N. Pal. (Eds.), *From Strategy to Execution*, (pp. 65–86). Berlin: Springer.

[50] Nidumolu, R., Prahalad, C., and Rangaswami, M., 2009. Why sustainability is now the key driver of innovation, *Harvard Business Review*, 87, 56–64.

[51] Lee, K., and Casalegno, F., 2010. An explorative study for business models for sustainability, *PACIS 2010 Proceedings*, Taipei, Taiwan, 47.

[52] Braungart, M., Mcdonough, W., and Bollinger, A., 2007. Cradle-to-cradle design: Creating healthy emissions – A strategy for eco-effective product and system design, *Journal of Cleaner Production*, 15, 1337–1348.

[53] Kumar, K., 2004. *From Post-industrial to Post-modern Society: New Theories of the Contemporary World*, 2nd ed. Hoboken: Wiley-Blackwell.

[54] Li, Y., Xue, D., and Peihua, G., 2008. Design for product adaptability, *Concurrent Engineering*, 16, 221–232.

[55] Woolley, E., Sheldrick, L., Arinez, J., and Rahimifard, R., 2013. Extending the boundaries of energy management for assessing manufacturing business Strategies, *Proceedings of the 11th Global Conference on Sustainable Manufacturing*, Berlin, Germany.

[56] Seow, Y., Rahimifard, S., and Woolley, E., 2013. Simulation of energy consumption in the manufacture of a product, *International Journal of Computer Integrated Manufacturing*, 26 (7), 663–680.

Permissions

List of Contributors

Fernando López-Valdez, Carolina Pérez-Morales and Mariana Miranda-Arámbula
Agricultural Biotechnology Group, Research Centre for Applied Biotechnology, Instituto Politécnico Nacional, Tlaxcala, Mexico

Fabián Fernández-Luqueño
Sustainability of Natural Resources and Energy Program, Cinvestav-Saltillo, Saltillo, C.P. Coahuila, Mexico

Alden D. Smartt, Kristofor R. Brye and Richard J. Norman
Department of Crop, Soil, and Environmental Sciences, University of Arkansas, Fayetteville, USA

Kakouei Aliakbar, Vatani Ali, Rasaei Mohammadreza and Azin Reza
Chemical Engineering Department, College of Engineering, University of Tehran, Tehran, Iran

Azin Reza
Department of Petroleum Engineering, Faculty of Petroleum, Gas and Petrochemical Engineering, Persian Gulf University, Bushehr, Iran

Francesco Fantozzi and Pietro Bartocci
Department of Engineering, University of Perugia, Perugia, Italy

Pradeep Kumar Malik, Atul Purushottam Kolte, Arindam Dhali, Veerasamy Sejian, Govindasamy Thirumalaisamy and Raghavendra Bhatta
ICAR-National Institute of Animal Nutrition and Physiology, Bangalore, India

Rajan Gupta
Indian Council of Agricultural Research, New Delhi, India

Adrian Ioana and Augustin Semenescu
University Politehnica of Bucharest, Bucharest, Romania

Zuzana Jelínková, Jan Moudrý Jr, Jan Moudrý, Marek Kopecký and Jaroslav Bernas
University of South Bohemia in České Budějovice, Faculty of Agriculture, České Budějovice, Czech Republic

Donatella Porrini
Universita`del Salento, Lecce, Italia

Elliot Woolley and Shahin Rahimifard
Centre for Sustainable Manufacturing and Recycling Technologies (SMART), Wolfson School of Mechanical and Manufacturing Engineering, Loughborough University, Loughborough, UK

Yingying Seow
Jacobs, New City Court, London, UK

Jorge Arinez
Manufacturing Systems Research Lab, General Motors R&D Center, Michigan, USA

Index

www.ingramcontent.com/pod-product-compliance
Lightning Source LLC
Chambersburg PA
CBHW062004190326
41458CB00009B/2964